THE EVOLUTIONING OF CREATION

AN ALTERNATIVE VIEW OF MODERN COSMOLOGY

Volume 2

Stephen J. Bauer

INTRODUCTION to SJBauer's Theories of Universal Significance

ISBN: Softcover 978-1-4628-8759-0

This book was printed in the United States of America.

To order additional copies of this book, contact:
Xlibris Corporation
1-888-795-4274
www.Xlibris.com
Orders@Xlibris.com

TABLE OF CONTENTS

DEDICATION

To my readers who I empower to continue my search for the answers to life, existence, and universal knowledge

and

To my family that they might understand my point of view

INTRODUCTION

"Precepts of General Relativity"

I would like to begin where I left off, by further ingratiating my view of Space-Time analysis within the accepted precepts of General Relativity. As I can not espouse to any detailed explanation of these precepts, I intend to paraphrase my understanding of these concepts into an oversimplification of their accepted beliefs. I relate them as beliefs because, similar to non-secular presentations of the universe, scientists are divided into differing divisions of thought on the subject of cosmology; each built upon their own beliefs. Among such beliefs their have been many epiphanies and prophets to be sure. The more well-known of these cosmological theories are the *Static Universe* introduced by Albert Einstein, the *Oscillatory Universe* presented by Richard Tolman, the *Big Bang Theory* of the universe proposed by Georges Lemaitre, the transitional expanding universe theory promoted by Fred Hoyle as the *Hot Big Bang Theory*, the *Steady State Theory* postulated by Thomas Gold and Hermann Bondi and also Fred Hoyle, and finally the *Inflationary Universe Theory* formulated by Alan Guth and Andrei Linde and Paul Steinhardt and Andy Albrecht.

Subsequently a consensus for modeling cosmology was agreed upon based on the work of four scientists: Alexander Friedmann, Georges Lemaître, Howard Percy Robertson, and Arthur Geoffrey Walker. Occasionally referred to as the FLRW, FRW, FL, or RW (e.g., a complete or partial combination of their last initials) Universe, it presents a metric used to explain Einstein's field equation of general relativity and thus became the foundation for the currently understood version of the standard 'Big Bang Theory'. And yet outside of this consensus, there still persists a variety of alternative theories that are either likened to a pseudo steady state theory or a varied flavor of the big bang theory. However my theory differs from all of these in the fundamental presentiment of an evolving Space-Time continuum relative to the dimensional interactions of Space and Time by degrees; i.e., minimum and maximum.

In understanding the accepted standard model of the 'Big Bang', one must believe in the constancy of certain variables in the universe; i.e., the speed of light (SOL). The 'Big Bang' purports to be a model of an open universe that is close to being spatially flat, thereby allowing for an infinite expansion. However anything less than a flat universe, by all calculations, would eventually expand toward a spherical shape along its own Space-Time curvature: ergo, a closed universe. Thus it would support a finite expansion and an eventual contraction of the universe. This contraction is sometimes referred to as the 'Big Crunch'. Thus, therein lays this complementary prediction of a more semi-evolving universe per the 'Big Bang' model. Herein belies its proclivity toward a model that portends the constancy of its geometric variables. I describe this model of the universe as 'semi-evolving' because it does present a beginning and an end developed upon the assumption of these constants of velocity in its homogeneous and isotropic expansion, and resulting contraction.

It is this very assumption of velocity constants and its corresponding constancies in acceleration, presaging a more homogeneous universe, by which my model would also differ. For trapped in Time and Space are the logical reckonings of the more discrete aspects for its heterogeneous nature. Therein this widely accepted concept of an expanding universe per the 'Big Bang' model differs from my concept of a fully evolving Space-Time continuum per my alternative model; i.e., in the semi-evolving universe, the cosmological model considers the universe as more uniformly whole per its expansion. Specifically a semi-evolving universe is consistently and uniformly expanding at a constant rate over the life cycle of its being; i.e., the Space-Time continuum of the universe provides for the one-time evolution of being from birth to death, from 'Big Bang' to 'Big Crunch'. Alternatively a more evolving Space-Time continuum provides for a multifaceted universe of ongoing dimensional evolutions in differing degrees of Space-Time, in which the universe's considered expansion is a result of its sundry dimensional conditions and transitions. Via a more multi-faceted evolution, my alternative cosmological model of the universe, Time is consistent but not constant and Time is not just a supplementary dimensional extension of Space. It is this consistency of a multidimensional Time that provides for an ever accelerating universe as the motivation for its being, rather than just the projection of its inception toward termination.

Indeed, what is really meant by an expanding universe? This is not a simple concept. And while these inflationary theories provide a variety of abstractions by which to visualize there intent, the 'Big Bang' expansion promotes a more ill-considered fourth dimensional model of an explosion from a singular point of origin; i.e., as what might possibly be misconstrued from an implication of the term 'Big Bang'. Rather the 'Big Bang' visualization is really the abstraction of a more even distribution of expanding matter, or spacial sections, upon which our measure of the universe is contrived. As demonstrative of this concept, from the perspective of an inflationary universe, an analogous measure of these spatial sections can be abstracted as balloons all inflating at the same rate of expansion within an infinite unknown void or abyss. To further expound upon this visualization, imagine that all balloons are touching and that in the center of each inflating balloon there are observers. Subsequently each of these observers could then observe that every other observer was moving away from them as the balloons equally and simultaneously inflated their spacial sections. Noted in this analogy is that only the spacial sections are growing in size and not the observers; for if everything is expanding at the same rate, then nothing is measurably expanding.

Even still, the 'Big Bang' describes more than just an expanding universe; for it also represents a semi-evolving universe as well: Hence over time its perspective, within this more traditional consideration of four dimensional (three of Space and one of Time), also transforms. To this end there has been ascribed a commencement force for this expansion, explained as that hot dense plasma whose heat fueled its prevailing motivation for inflation within the medium of some unknown void. An analogous representation would also be akin to boiling rice, in which each spatial section was uniformly expanded due to this commencement force. Complementarily a termination force is related relative to a cooling down of this hot universe, in which each spacial section will be uniformly contracted toward a new singularity. This imparts a more corollary concept giving way to a contracting universe: The expectation that there is a symmetrical de-evolving process by which all matter can return to its original state.

So then what is meant by a contracting universe? This is not necessarily a simple concept either. Indeed, it is not just the ill-considered fourth dimensional model of an implosion from a voluminous universe, as what might possibly be misconstrued from an implication of the term 'Big Crunch'. Rather the 'Big Crunch' is really the more even distribution of contracting matter, or spacial sections, which makes up our measure of the universe. As demonstrative of this concept from the perspective of a contracting universe, an analogous measure of these spatial sections can be abstracted as balloons all deflating at the same rate of expansion within an infinite unknown void or abyss. To further expound upon this visualization, imagine that all balloons remain touching and that in the center of each of these deflating balloons there are observers. Subsequently these observers could each then observe that every other observer was moving closer to them as the balloons equally and simultaneously deflated their spacial sections. Again noted in this analogy is that only the spacial sections are shrinking in size and not the observers; for if everything is contracting at the same rate, then nothing is measurably contracting.

Yet an alternative version of such an eventuality could be visualized quite differently, and provide equally empirical results. Imagine instead that all these fully inflated balloons have established themselves within the universe such that their centers remained fixed. To further expound upon this visualization, imagine each balloon as a galactic boundary, and these balloons no longer need to remain touching while they shrink. Now as each balloon deflates equally, the observers shrink along within their shrinking galaxies. The observers would then be able to measure the illusion that each galaxy is moving away from them. Subsequently these observers could each then observe that every other observer was moving away from them as the balloons equally and simultaneously deflated. Could it be that everything we believe we know about the constancy of mass measure be truly inviolate of its dimensional size? While I am not purporting this model as valid, it does provide some interesting ramifications.

Still this ideates yet another professed constant of the 'Big Bang' model: There is a set amount of matter and energy in the universe that is manipulated via this constant semi-evolution toward expansion. While all these constants are great for calculations, they present more problems than they resolve. For example, what factors are used to decide the amount of matter and energy that would have been provided at the time of the universe's creation? Are there other such universes that may have been afforded more or less matter and energy than ours? If the universe is cooling, then why is all matter still accelerating; shouldn't our universe be slowing down as it cools? If the 'Big Bang' is to be followed by a 'Big Crunch', will another 'Big Bang' be expected to follow? What are the implications of multiple instances of our universe, e.g., multiple 'Big Bangs', or even multiple universes in the scheme of cosmology?

Although the 'Big Bang' model has its problems, these issues pale in comparison to the ill-conceived notions purported by the first 'Steady State' models. The first 'Steady State' models tried to infer a non-evolving, or static, condition; probably building upon Einstein's earlier view of a static universe. The 'Steady State' theorists asserted that although the universe is expanding, it nevertheless does not evolve over all of time. Rather like the concept of the paleogeographic animation of a single Pangaea, its intent was to suggest that the universe did not require a beginning or an end. This theory ideated that the universe was little more than a container for cosmological events in which pieces were shuffled, reorganized, and even recycled. This short-sightedness doomed this model upon the discovery of cosmic background blackbody (CMB) radiation. For the discovery of cosmic microwave background radiation was considered proof of a more evolving and dynamic universe, and a validation of the 'Big Bang' model. For the Steady State model to persist, it now had to account for the notion of ever-generating CMBs as a condition of their ideal for continuous existence. Later, the 'Steady State' model would be revised to include the notion of a more Time altering, or evolving universe, but the damage of its first inception was already done.

To understand this discrepancy, one needs to understand the accepted perspective of the Space-Time continuum as supported via the 'Big Bang' model. The 'Big Bang' model is predicated upon the assumption that the Space-Time continuum represents a constant dynamic condition. Subsequently, since the velocity of light is considered the ultimate constant, light speed can be used as the basis for all measurements. Thus the 'Big Bang' model purports that this observable, emitted cosmic background radiation is a perspective window into our universe's previous condition; by virtue of an age measurement based upon the shifted wavelength of light speed. Light speed age is the conceptual measure of a photon's shifted wavelength. Such is it that by measuring the shift of a radiated wavelength, the extent of this shift can be used to calculate how far into the past a previously radiated event once existed.

Another aspect of this wavelength shift demonstrates a unique property of direction. While a red-shifted wavelength indicates a motion of accelerating distance away from an observer, a blue-shifted wavelength indicates a motion of a decelerating distance toward an observer. In effect, a red-shifted wavelength represents the stretching of light, characterized by the wavelength's dilatation over time, while the blue-shifted wavelength reflects more of a compression of light, characterized by the wavelength's contraction over time. Yet whether the radiated event is red-shifted or blue-shifted, all received wavelengths are still the measures of relative motion either from or to an observer. Therefore, where Space-Time can be considered to support the velocity of light as a constant, these same wavelengths can become a measure of distance to and from the observer as well. And as such, these perceived wavelengths are merely the signature of received radiations for past events. Subsequently these CMBs were then interpreted as the signatures of hot dense plasma that pervaded in the existence of our early universe, and whose motion demonstrates the continual expansion of a more dynamic universe.

Further exploring this constancy of light speed for such measurements provides insight into how the 'Big Bang' model gained favor over other cosmological models. For example, within our own solar system, the light rays traveling from the Sun are presented to an observer on Earth eight minutes after leaving the surface of the Sun. This lag in time is representative of the approximated eight minutes it takes for the Sun's rays, or streaming light particles, to reach Earth at the speed of light. Yet we are relatively close to the Sun when considering it would take a little more than 4 hours for light reflected from Neptune to bounce back to Earth. To expound upon this imagery even more, let's imagine two separate immortal observers within our own galaxy. These two separate immortal observers at either end of the Milky Way galaxy would only observe the Past radiated events from 100,000 years prior to either's Present-Time, based again on the constant speed of light for measure. And if either would attempt to contact the other, they could not expect an answer for over 200,000 years. This is longer than most believe that there was even a species of mankind on the planet Earth.

However it has been determined that the nearest solar system to our own, that might support a similar life environment to that of Earth, is only about 130 light years away in the constellation of Pegasus. If one is truly to believe in the 'Big Bang' model of a homogeneous and isotropic pattern in the evolution of the universe, then intelligent life on other habitable planets could only be expected to have evolved within about the same time frame as ours, give or take 800 million to 13.5 billion years. Yes, the birth of our own galaxy the Milky Way is still in question and has only been narrowed down to an estimate sometime between 800 million to 13.5 billion years, depending of your source for such information. As a side note, suspiciously enough the calculated age of the universe is not much more than the most extreme estimated age of our own Milky Way galaxy. This coincidence suggested to me that such calculations for the entire universe might merely be based on only our little corner of the universe and not relative to the overall universe at large. This being the case, however, the first contact for any Search for Extraterrestrial Intelligence (SETI) could be expected at any time, or may have come and gone billions of years prior to the existence of our own solar system, approximately 4.5 billion years ago.

Still, the detection of extraterrestrial intelligence is based solely on our definition of what humans consider to be intelligent life. There is a flawed logic that the measure for extraterrestrial intelligence should be based on their ability to inadvertently or purposefully transmit the presence. This measurement must take into account that a discernibly unique energy signature is resonated from some form of technology similar to our own. I believe had other intelligent beings evolved much earlier than life on Earth, and had they the inclination to purposely send out a signal for other life beyond their own solar system, that their signal would have been delayed, masked, perhaps gravitationally distorted, or even lost to a rogue black hole. And should a line of such expensively long distance communication be established, at best we could swap theories of the universe, or discuss societal problems, or religious and cultural philosophies. Neither inhabitants would be more better off than the already are, because being extraterrestrial or even living longer doesn't make one more intelligent.

Similarly, if one is intending to find a suitable planet for colonization, then they should not base their decision on what they observe from Earth due to this time measured delay. Rather they would need to consider how the candidate planet might have evolved prior their expectant arrival in that planet's future; for only the planet's past condition is ever observable. As our nearest neighboring exoplanet candidate is over 130 light years away, it would still take our fastest spaceship some 2 million years to reach this planetary system. And while considered light speed travel may be explored to resolve this travel barrier, it would still propose a time lag of 130 years for a probe to reach the candidate planet and an additional 130 years to receive any report back from a probe (providing nothing went wrong). Not to belabor the point, but our science and technology are light years from enabling such an endeavor; but I digress.

So as the 'Big Bang' model presents a perspective by which the universe is expanding and evolving using a measure of past radiated events, some may be led to believe that we, the human race, reside at the pinnacle of this universe's evolution; for there is no measure to observe evidence of any future radiated events. Rather, in examining this notion of a dynamically evolving universe, one can only observe signal evidence of past events. Considering this perspective of the 'Big Bang' model, then anywhere in our defined universe an observer would only receive these similarly shifted radiations; for each homogeneous spacial section is expected to have evolved in a similar fashion. And each homogeneous spacial section would be similarly measuring their condition within this expanding universe.

For example: While these CMBs may represent the primordial condition of our universe, they are only representative of past radiated events. Therefore at the extreme point in Space-Time to which an observer can view these CMBs, an opposing observer from this extreme point could be looking back at us and only see these CMBs as well. This would also imply that the universe is much larger than we are able to measure; for at the extreme point in Space-Time to which our first observer can view these CMBs, a second observer at this extreme point can see even further into the universe than the first observer.

One might consider that both events, the Space-Time perspective of a co-located Present-Time and Past-Time, exist relative to the observer and thereby provide a validation of the earlier revised version of the 'Steady State' model. It's all an illusion of our perspective for measure. Still while the 'Steady State' model was an attempt to provide a condition by which our journey through an infinite universe could be theoretically abstracted, the 'Big Bang' shifted our focus toward a more measurable, time-based universe. Yet regardless of which cosmological model is employed, it does point out that there is no real way to observe any Future-Time events from anywhere in the universe; or even another Present-Time event, for that matter. All measureable input is filter through a time delay from the event to the observer.

Still if red-shifted wavelengths represent a view of our past, would it be plausible to expect that blue-shifted wavelengths could represent a view of our future? The answer is "no", because, even as the radiated event is blue-shifted, it is still a distance from the observer. Although not previously discussed, both red-shifted and blue-shifted events are only the signature of energy expended, either away from or toward an observer respectively. These shifted wavelengths can only provide varying degrees of Past-Time events. If fact, even though blue-shifted wavelengths are accelerating toward an observer, their measure is still akin to that of red-shifted wavelengths because all matter in the universe is expanding accordingly to the 'Big Bang' model. However, in the case of a blue-shifted event, each succeeding event presents a perspective toward a more Present-Time rather than a more Past-Time. Consequently one can only observe the emanating signature of energy expended whether traveling toward or away from an observer, respectively blue shifted or red shifted.

Instead, much like the desperate radiated events lost within a 'Black Hole' in Space and Time, a measure of Future-Time events can not be observed within the confines of our dimensional existence. Just as a 'Black Hole' provides a Space-Time continuum by which such radiated events are accelerated beyond our measured condition of relativity in Time and Space, so too the radiated events of Future-Time can not be measured. In fact, such radiated events within Future-Time represent a measureless variance in relativity that can never be approached or encounter. These radiated events, which have evolved to become Future-Time events, provide no measure of kinetic consequence within our dimensional frame of reference. From the point of view of the observer, the lost energy signature has evolved into a more potential event. From the point of view of this lost event, it has been dispersed upon the cosmic background of its renewed relativity (but I should not get ahead of myself). Thus, one can only ever observe or ever prognosticate upon Past-Time, evolutionary events from the relative perspective of their own Present-Time, evolved condition. The travesty of real Time is that it is not a dimensional subsistence which can be traversed. Time can only be mirrored to reflect prior time events; Past-Time is only relative to the Present-Time observer, or collection device, as the arbiter of such input.

Subsequently the precepts of relativity are every evolving upon our understanding of their evolutionary events. Therefore I would like to introduce my equally contentious alternative theory in which a continuous creation and a continuous termination can coexist in balance upon an ever evolving Space-Time continuum that is not Time constant, but merely Time consistent. Instead of considering only the observable presentation of the cosmos in shifted wavelengths, I would like to abstract beyond an event's ability to be observed and measured. I would like to do this to convey a greater understanding of the universe that is so much larger than what we can ever imagine within these measured limitations. I would like to abstractly demonstrate a universe that exists via the dynamic environment of its varying dimensional conditions. I would like to propose a universe that is motivated via its evolution from 'minimum Space' and 'maximum Time' to 'maximum Space' and 'minimum Time'. I would then like to infer a universal condition whose very being is predicated upon a balance of an existence from and within nonexistence; where nonexistence represents the considered unknown void or abyss. My vision is of a universe in which 'all Time' and 'all Space' reside in dimensional juxtaposition to each other. It is a universe whose existence is merely balanced upon the relative stability of energy for matter within this ever altering dimensional juxtaposition.

AUTHOR'S NOTE

From the perspective of this alternative model, this alternative cosmological theory, I can only abstract a presentation of this paradigm; for one can only observe that which can be manifested within their own Space-Time relativity. Similarly as one can not perceive Future-Time events, like the terminus of existence, one can not observe all aspects of these varied dimensional Space-Time continuums, like the creation of existence. In example, the being of an observer within the traditional understood 4th dimensional condition could not perceive the being of an observer within an extractable 3rd or even 2nd dimensional condition, and vice-versa.

Considering the concept of dimensional conditions, or frames of reference, this abstracted paradigm requires a better purview of existence beyond the simplistic view of a traditional fourth dimensional Space-Time continuum; i.e., three dimensions of Space and singular presentment of all spacial punctuation in Time. Instead, there should also be multidimensional variations of Time, as well, rather than just of a multidimensional Space in Time. Think of it as similarly abstracted of Space: s/t demonstrates a linear constant of one direction in Time, s^2/t^2 engenders a more progressive constant maintaining Time in two directions, and s^3/t^3 enlivens a constant consistency in three directions of Time. Maintaining Time in two directions, by measure, prepossesses the bounding of infinite Time within the curvature of its persistence. Whereas maintaining Time in three directions, by measure, prepossesses a bounding of acceleration or deceleration within the completion of its own measured volume.

Still even here, within three dimensional Time frames, the expression of gravitational acceleration is not enlivened in an amalgamation for the spacial preponderance required of the creation for mass. However, furthering this notion of such dimensional variations for Time, s^4/t^4 does embrace the notion of Time in four directions. By maintaining Space and Time in four directions, by measure, it prepossesses a bounding of (acceleration or deceleration)/time within a vortex of its own impetus. Subsequently, the gravitational force (acceleration or deceleration) is a consequence of eight dimensions: four dimension of Time imbibing a complementary four dimensions of Space. A presentiment for the actual model in four dimensions of Space should be: points for line, lines for plane, planes for volume, and volumes for density (representative of the separation of positive and negative density for mass).

In eighth dimensional events (understood as traditional fourth dimensional existence), the dominant Space-Time relationships are warped to allow for matter to form into mass. Herein, Time is internalized within matter as energy, required to maintain a mass' density. Ultimately, matter accumulates Time, as this internalization renders less Time to be available within the relativity of its Space-Time continuum. This continual internalization causes a kind of temporal tension, or the degree to which Time is warped relative to its surrounding Space. Subsequently, matter ages as it accumulates more and more Time relative to its Space. In the end, the surrounding Space becomes devoid of Time causing an imbalance that must be corrected. It is a unique perspective, to be sure, that would require some retraining of the mind.

Chapter 6

"The How of Evolution in Creation"

Prologue

How?... a subsequence of the interrogative 'Why,' implies a convention of purpose towards expression. In this dissertation, its purpose is to convey the 'How' of it all; to propose a recognizable cosmogony unto an expression of the universe as it exists today. Whereas 'Why' implied the intention of such a purpose, as presented in the first volume of this compendium, the reader was left with the question, "How does it all work?" Wherein this newly introduced paradigm of cosmogony, as first captured in volume one, has identified some new puzzle pieces with which to work, how does one logically assemble these new pieces of the puzzle for a proper fit? Truly there are many methods and models by which cosmography is viewed and influenced as to the 'how' one should constructs these puzzle pieces. If one measures these pieces as traditional fourth dimensional quantum particles, they will view its construction differently than if they were to relate these pieces as one-dimensional superstrings.

Whereas quantum particle physics tries to draw order from within a traditional fourth dimensional perspective of the universe, the quantum field of string physics attempts to extract an order from within a more multi-dimensional perspective of the universe. Indeed, a consistent quantum field theory of one dimensional superstrings can be explained using 10 Space-Time dimensions. Seemingly, the string theory purports a stretching of particle energy rather like the imagined accelerated condition of mass traveling into a black hole. Certainly, this extra dimensional perspective also provides an extra measure of mathematical possibilities with which to define the interrelations and structures of the universe. However apodictic these mathematical proofs might seem, for these proofs to be accepted as scientific fact they would still require reproducible corroboration via some empirical evidence or physically observable characteristics within a traditional fourth dimensional context. An example of how these methods, by which one may fabricate an understanding of the universe, influence how these puzzle pieces should fit together is in the limitations of their imposed measures. One considered conclusion of this interrogation is that particle theorists commence with a firmly supported Big Bang model of the traditional fourth dimensional perspective of the universe, while superstring theorists can begin with a more evolving pre-Big Bang dimensional manipulation of the Space-Time continuum.

In remanding the original conflict of attitude, as discussed in the first volume of this compendium that beleaguered the 'why' of it all, one observes its continuation in a new guise. This conflict of attitude now begets a beleaguerment unto the 'how' of it all. In this seemingly irreconcilable quest for the answer to the question of what came first, Creation or Evolution, one is tormented by the demons of their own devise. Yet, exploring such idealisms really all depends on how one employs or qualifies a notion of Creation. Particle theorists might view Creation as the first reconciliation of matter for mass. Superstring theorists, on the other hand, might view Creation as more of an apriori revivication from a dimensional aspection in the convergence of Space and Time for the engendering of a potential for matter.

And still others, as theosophy would profess, consider Creation as the instantaneous appearance of all apperceptive life that is exhaled, expelled, and/or evaginated from a singular unified entity. More or less, particle theorists are in agreement with this theosophists' conception of Creation for the universe, except for the theosophists' assertion of expecting instantaneous life. Of the latter, the aberrant nature of the term 'instantaneous' in this theosophic reference, seeks to qualify the appetency of Time in Space while ignoring any foundation for its being. From this theosophic assertion, it is then an apostasy to apply an evolutionary convergence of Time and Space with which to qualify this dimensional affinity. In this single-mindedness, the premise of such theosophies supports more the original, but short-sighted, Steady State model of the universe; i.e., that which 'is,' has been and always will be unto the commencement of Time and Space.

However, if there is to be a dimensional aspection in which the notion of Time converges with an ideal of Space, then there must be a dimensional aspection wherein such facets of relativity are yet dispersed. And if this is to be properly idealized, then there must also be a dimensional aspect of nonexistence relegated as the cosmological apogee wherein there is no convergence of Time and Space. Ergo, without a convergence of Time with Space, the relative logic to the understanding of a notion of Creation is that it can 'not be' before it is 'to be.' Yet from this condition of 'not being' that precedes Creation, there must also be an evolutionary state or condition from which this dispersion in 'no Time' and 'no Space' can be abstractly imagined and/or idealized toward a convergence in 'all Time' and 'all Space'. Therefore it is in this evolutionary insistence of existence from nonexistence in which a state of Creation is revivified for the reality of our universal condition: Cosmogony toward Cosmology. It is then from within a condition of 'no Time' and 'no Space', or nonexistence, that a reality for Creation is imparted to regulate the fabrication of our universal condition.

Still, discerning the 'how' of it all is invariably a problem of some complexity for most everyone. Some just have difficulty with abstractive contemplation, while others regulate their thoughts upon the supplanted knowledge of previously pre-accepted affirmations or ideologies about our universal condition. And yet, there remains an underlying framework for the dimensional aspection that influences purpose toward an expression in the aberrantly affinitive nature of the Space-Time continuum. While two such contentious pillars of previously pre-accepted truth and knowledge in our society, theology and science, battle for a foothold in our consciousness, this unique dimensional aspection is all but ignored.

As these two philosophies were previously encountered in a quest for the 'why' of it all (in the first volume of this compendium), these groups hold the preeminently authoritative positions of qualifying reality into facts relative to their own campaign of idealizations. Each campaign then splinters into their own varied schools of ratiocination when approaching promulgation on the subject. And each of these camps, having spent a number of lifetimes building up and tearing down countless renditions of their own impressions of cosmogony, splinter yet again in revisiting revisions in support of their promoted dogma. As if this were not enough to intimidate or discourage the average individual from thinking 'outside-of-the-box,' each splinter group is forever vigilant and protective of their idiosyncratic pragmatism; as if they believed their veracity (or maybe their importance) would be impugned, insulted, or threatened upon the introduction of any new paradigm which has not been firmly indentured within the existing heritage of their supplanted knowledge base. Indeed there are specific terms and concepts by which one may communicate and compare such ideologies, and there needs to be common ground, but the 'how' of combining and fitting these puzzled pieces should always be allowed as fuel for debate.

Such are the winds of conventional wisdom, which change from one era of discovery to the next. What is considered the fact of one age can become the fantasy of another. And what is considered the fantasy of one era may later become the fact of another. However, if one is to expand upon the horizons of such abstractive expressions toward an understanding of the ultimate truth in a determination to the 'how' of it all, then new paradigms should to be ideated and explored. For although established facts and formulas are needed to mediate between the myriad of theories and abstractions that abound, these theories and abstractions, by themselves, should not be the imposed boundary for establishing facts. Subsequently, these checks and balances should not dissuade new views of the universal condition. Rather all these sundry theories and abstractions of our universal condition are quite subjective and personal. They have been and should remain the realm of all thoughtful dreamers with an aptitude for analytical design. Choosing one's own choice of ideology, whether secular or non-secular, for one's own theoretical existence is as important as exploring one's own identity or choosing one's own ideality. Yet, although one needs to consider existing ideologies for guidance, in the end each individual should take advantage of their own free will to create their own version of reality; for that's what living is all about.

Understanding Universal Boundaries

In approaching the 'how it all works' in terms of one's universal existence, one needs to consider the physical phenomenon for the formation of matter itself. As I have previously presented and discussed in the first volume of this compendium, Time and Space conform for a semblance of dimensional convergence that bounds a continuum for existence from within an absolution of non-convergence expressing nonexistence. The presentment of this paradigm provides for an evolutionary progression in the adjunct relativity of dimensional interweaving as the action of a temporal cosmogonic transience prevailing in harmony with its preternaturalness. What does this all mean? The formation of matter is more than just its facilitation of its condition within the dimensional constants of Time and Space. The formation of matter is the resultant state of considered degrees in the dimensional fabric of Time and Space; i.e., 'no Time', 'all Space', 'all Time', 'no Space', 'minimum Time', 'minimum Space', 'maximum Time', 'maximum Space', etc. It is these degrees of abstraction that allow for the concinnous comingling of distinct continuums of relativity. And it is these relative distinctions that facilitate a temporal cosmogonic transience for evolving realities within the relativity of this alternative paradigm. In this, our temporal cosmogonic transience, the universe can be expressed as the extrinsic release of existence within the dimensional opponency that ripples through a perpetual cycle of temporal absolution.

But how to assemble these abstracted degrees in the dimensional fabric of Time and Space; that is the real challenge. Like any puzzle, the abstraction of the enigma plays an important role in understanding how it all fits together. Concepts like 'no Time' or 'no Space' represent ideals of infinity just a much as concepts like 'all Time' or 'all Space.' Any combination of 'no Time' or 'no Space' results in nonexistence. Still, at what point are concepts like 'all Time' or 'all Space' relevant to our current relative existence. Truly the acquisition of 'all Time' and 'no Space' is the prerequisite progenitor of creation; related via theorists as the instaneous presentation of existence. Conversely, the evolutionary progression of an accelerating universe tends to minimize the dimensional influence of Time and eventually nullifies its existence. In a universe that is ever-beginning and ever-ending, it can also be said that the universe is never-beginning and never-ending. It is this ripple in the dimensional opponency of Time and Space for evolving realities that inspirits spatial continuums with motivation and direction; engendering the transience of an ever changing, ever temporal Present-Time, Space-Time continuum.

Present-Time events, seemingly separate from the extrinsic boundaries of Past-Time events or Future-Time events, form the temporal anomaly that divines the relativity, or relative condition, of our Space-Time continuum within this evolutionary progression. Therefore, within the expectations of our considered universal condition, the direction of our evolving realities would seem to be rationalized from a dimensional convergence of 'maximum Time' and 'minimum Space' toward a dimensional convergence of 'minimum Time' and 'maximum Space'. Where upon this rationalization in which the acquisition of 'all Time' and 'no Space' is the considered entry point for this universal condition, then the acquisition of 'no Time' and 'all Space' would be the considered exit point for this universal condition.

Such spacial continuums are rather likened to the 'Big Bang' abstraction of a more even distribution of expanding matter, or spacial sections, upon which our measure of the universe is contrived. However as a consequence of our varying dimensional condition, these spatial sections of our relative existence are accelerated along from a dimensional juxtaposition of 'minimum Space' and 'maximum Time' to 'maximum Space' and 'minimum Time'. Thereupon the relativity of these spatial sections will ever continue to evolve as well. It is this Space-Time progression that both predicates and precludes the continuance of our Present-Time relativity, while maintaining the evolving continuance of its being. Thus, as each spacial section is evolved along this progression within Space and Time, a new spacial section replaces its former relative dimensional condition. Somewhat like the wire framed conditions imposed on soap bubbles, the addition of each new relative dimensional boundary in the wire frame alters the shape of the soap bubble. Similarly, upon each alteration in the dimensional boundary of our Space-Time continuum, the relativity of our existence evolves beyond it previous condition; separate and unique of the next and previous ongoing Space-Time events.

Considering that one can not view the ongoing events beyond the perspective of their own relativity, it is easy to similarly understand why one can not observe the Present-Time events of a distant radiation's origin. In this way they can also not view the Present-Time condition of the universe's origin or the Present-Time condition on the termination, because these are ongoing events that can only coexist in dimensional conditions different from our own current relativity within the universe. Rather all that can be observed or measured is only within the relative condition of the cosmos from the perspective of our Present-Time, Space-Time continuum. And the Present-Time, Space-Time continuum of our spatial section is only relative to its existence within its own varied dimensional juxtaposition. In fact, if there were Past-Time events of newly created matter or energy upon the universe's origin, we would not be able to view these events not only because of their imposed time dilation, as unperceivable beyond the shifted wavelengths of the cosmic background blackbody (CMB), but also because they exist in a differing dimensional frame of reference. More interesting is that such creation events would have preceded the very existence of light signatures.

Another consideration is the consistency of Time. Each observable Past-Time event remains relative to an observer's Present-Time event. While their 'clocks' may be different, their 'clocks' remain consistent in relation to one another. A Present-Time observer can no more view a preceding event to the Past-Time event input that they have received from a distant point in the cosmos, than they can view their own preceding event to the Present-Time event in which they are currently occupying. While careful study will allow an observer to view the succeeding Past-Time event inputs from that distant point in the cosmos, it can never truly observe any of that distant point's preceding evolutionary events. It is only ever just the one point in time to which they are relative to each other upon the consistency of their 'clocks' that will ever be observable or measurable. Like a recorded movie that has aged, the presence of these past radiated events is only a distorted record of the original event. There is no real availability of this spacial section's past relative condition in the current Space-Time frame of reference, frame of existence; each spacial section can only exist in its Present-Time relative to their condition in the universe. These past radiated events are just the archeological artifacts of our viewable cosmography within our portion of the universe. Thereupon the evolution of our Space-Time condition within the universe renders any reference to our previous condition as nonexistent. Subsequently, it is this consequence, in the varying dimensional transitions of our dynamic environment, which precludes the ability to travel into the relativity of another Past-Time or even another Future-Time.

Consequently this thereby also precludes the consideration of traveling between differing continuums of Space-Time, liken to the abstract concept of traveling between differing continuums of Time (or Time Travel) within Space. Such is it that the being of the universe, as we have come to appreciate it, resides in the persistent balance of an existence from nonexistence; a relativity of consistence within the alterity of Space-Time rather than merely the condition of relative constants within a homogeneous Space-Time. Still for all practical purposes, the maintaining of multiple co-existing Space-Time continuums conflicts with the concept of energy-matter conservation within a homogeneous and isotropic pattern in the expansion (or subsequent evolution) of the universe. A primer of the supported model constant for the 'Big Bang' theory is that there is a set amount of matter and energy in the universe that is manipulated. Infusing more matter or energy into a Past-Time event would require a balanced consequence for a subsequent loss of matter and energy within the Present-Time event from which it was displaced. This would create a paradox in which the previous set amount of matter and energy in any given time frame is no longer constant and is no longer a relative consequence of its previous potential. In short, the rationalization of co-existing and/or competing Space-Time frames of reference undermines the ability for matter and energy to form or even exist.

The Epoch of Matter

Upon the event of Creation, the divining of Space in Time (or spaced Time) defines the first expression in, or being thereof, matter toward a material present. However this is matter in only an abstract sense of the ideal condition required to form the material from which the universe is now substantiated. This condition of matter is an evolutionary pertinacity that needs to be promulgated through more persuasive perturbations of the Space-Time continuum to procure a more evolved perspicuity in both the nature and promiscuity of its substantiation. Ideally, science defines this condition, in the personification of its pertinacity, as 'gravity.' Where upon its adverse condition, in the personification of its distraction, is defined as 'not gravity', or 'antigravity.' More recently, 'anti-gravity' has been distinguished by some scientists via of the more homogeneous term, 'dark-energy,' in hopes to distance its invention as separate of Albert Einstein's use of a cosmological constant. And yet 'gravity' is but another evolutionary digression in the temporal comingling of Space and Time.

Such is it that it is necessary to understand these scientific ideals so as to comprehend their relationship in matter and in the nature of 'how it all works.' Per the "Machining of Materiality" in volume one, the concept of gravity is the representative result of the opposition of existence to a state of nonexistence, while the concept of antigravity (or dark energy) is the representative result in the affinity of existence toward the state of nonexistence; i.e., respectively a more positive mass is attained via gravity versus a more negative mass incurred via dark energy.

Remanding an examination of the four fundamental forces, as previously ideated in the first volume of this compendium, maximum distance defines the very tabernacle of Creation in the abysm of nonexistence. Surely it is by far the most impalpable, impeccable, and immeasurable force of the four. From this ideal notion of distance, one realizes its relative antithesis in existence as resistance. Maximum acceleration more aptly expresses the ideal of streaming consistency, while realizing its relative antithesis in existence as constancy expressed via maximum deceleration. Thus these four fundamental forces are manifested relative to Time and Space as follows:

1. Maximum Acceleration (Minimum of Time - Minimum Deceleration)
2. Maximum Deceleration (Maximum of Time - Minimum Acceleration)
3. Maximum Distance (Maximum of Space - Minimum Resistance)
4. Maximum Resistance (Minimum of Space - Minimum Distance).

Distance personifies pertinacity as the veritable glue of existence, suffusing substance with an ideal and an identity in the purposeful expression of Space. The antithesis of resistance, distance abides as the infusion of inflection abetting the constancy of velocity; i.e., without distance there is no measure for velocity. Contra wise, maximum resistance is related as the confocal confluence of peragration abrogating maximum distance toward a focus in the ideal and an identity in the purposeful expression of Time; i.e., without resistance there is no constant for measure. Resistance to distance becomes the start point with which one can observe its maturity toward a precipitation in the corporal quality of nature as matter.

Such is the precipitous pre-conformity of this point of maximums that it personifies pertinacity as the veritable glue of existence. Matter abides as the infusion of inflection abetting the constancy of velocity to a being in palpable force; i.e., matter abides as bounded measures of Space in Time. Perceivable matter matures with increasing motion, relative to its Space-Time condition, into a view of existence as a form of energy. The ideation of matter necessitates an exploration into a view of its existence as the kinetic energy whose only barrier toward maximum acceleration (understood as maximum deceleration via a perspective from nonexistence) lies in its aversion to an introversion transition into potential energy; i.e., the end result of maximum bounded motion is no relative motion at all. It is this causal connection to nonexistence that is the progenitorship which distinguishes the residual resistance of Space from within the disparate dispersive influences of its Time; e.g., its potential 'to be' is that it 'is.'

Or another way of looking at it is that kinetic energy is the transition from a Past-Time event to a Present-Time event, wherein potential energy is the transition from a Present-Time event to a Future-Time event. Potential energy, from the perspective of non-existence, might also be understood as the kinetic embodiment of reference for origin from a Future-Time event to the next Past-Time event. Potential is the ability 'to be' based on what 'is.' Without potential energy there can be no kinetic energy; no consequential Time events for Space to exercise. Material subsistence abides via this affinity of existence to nonexistence in matter; e.g., its condition in the personification of its pertinacity as 'gravity' and its adverse condition in the personification of its distraction as 'antigravity', or 'dark energy.'

Matter is the catch-all term of physics for expressing the physical phenomenon of extensible existence. Whereas mass and energy have a whole host of well understood measurements, not everything is definable within these limitations. Thus matter is used to explain that which can not easily be defined: Matter is anything that occupies space and follows the laws of gravity and inertia. Definable mass is, more or less, the practical complement of gravity. As mass is used almost interchangeably with gravity, it becomes the measure of its comparison upon itself; i.e., mass is defined upon the arbitrary selection of one of its smallest detectable forms, the atom. However, for a more complete understanding of mass, mass is defined upon the apposition of its physical force or presence. Simply put, mass is compared against itself for measure.

And yet, although mass is the principle component of the gravitational equation, this use of a mass' measure alone can not account for the complete influence of gravity within the scheme of cosmology. The ratiocinations for these discrepancies are varied via such expostulated innovations as the notion of 'dark matter.' Dark matter is the concept of matter that does not contain atoms, or nonbaryonic matter. Dark matter is considered as the undetectable or 'missing mass' of the universe. In and of itself, this form of undetectable matter has no real mass with which to upset the current theory of General Relativity based on a gravitational constant. However, 'dark matter' remains as an expostulated innovation of fractional mass used to track these anomalous discrepancies of gravitational interactions. Indeed, the invention of dark matter as a distinct unseen force tends to enliven much fuel for debate. Perhaps dark matter is really a distinction in varying degrees of the non-density, or negative mass in matter.

In my opinion, the overall consideration for gravitational forces is more than just the measure of its relationship as a discrete mass to be translated via the predetermined 'gravitational constant' [e.g., G =(6.6742 ± 0.0010) × 10-11 {N m2 / kg2} = (6.6742 ± 0.0010) × 10-11 {m3 / kg s2}; where the value can also represented as ((2/3) ± 0.00010) × 10-10 {m3 / kg s2}, or approximately 1/15 billion]. Rather I believe that gravity might be better understood when considered via the combined synergistic influence of all matter. Obviously gravity is easily demonstrated upon the independent influences of discrete masses upon each other, but what of the synergistic influences adherent in the proximity of their groupings? In contemplating the gravitational attraction of galaxies or galactic clusters, does it really make sense to key in on just the largest object within such a grouping to account for the groups' overall gravitational force? How close together do two objects need to be to be considered as one? A good example of this synergistic influence in gravity is demonstrated upon a volume of plasma. Although the volume of plasma is considered as a discrete unit, it is actually made up of loosely bonded particles of ionized gas. Yet it has density and, thus, it has gravity. So even as the plasma is held together via gravity, the volume of the plasma exerts its own gravitational force relative to objects both outside and inside of its discrete unit. In fact, a star begins as a collapsing cloud of plasma composed primarily of hydrogen, with helium and trace amounts of heavier elements. Similarly one might view the gravitational influence of the whole Milky Way galaxy as either just a product of its centric 'Black Hole' or as the complete synergistic grouping of all mass within the Milky Way galaxy that induced its centric black hole.

Indeed, wherein gravity defines the condition of our universe, such an eclectic perspective toward the concern of its understanding is warranted to gain a better appreciation of its considered relativity within the scheme of cosmology. When taken as more than the iterations of its collective measurements towards its discrete mass relationships, this consideration fosters an extension of the existing understanding of our overall gravitational model. After all of its careful crafting, what really is this representation of a 'gravitational constant' other than the result of unknown factors in the measurement of discrete mass relationships? To understand the true nature of matter, one needs to understand the true nature of the overall gravitational force by which it is bound.

First I would like to approach an examination of the gravitational force as defined in the measured relationship of its discrete mass. In considering the true nature of the gravitational force within these discrete bodies of mass being measured, is its gravity really as uniform as its considered overall volume? The existing understanding of the gravitational model would appear to be based in an equation of a gravitational force on a more linear representation of mass. In example, a considered discrete body of mass is measured as if its density is equally distributed within its spherical volume. This linear representation can be similarly termed as 'geoid': e.g., the surface of equal gravitational potential upon a hypothetical ocean at rest which serves as the classical reference for all topographical features. From this representation, the further one travels into such a spherical volume of specific linearly distributed density, the less the traveler is affected by the gravitational force arrogate in the mass of the whole volume. In effect, when measured towards its center, the force affects of the mass' gravitational acceleration are considered abated. Finally, when measured at its center of a mass density, the affects of the mass' gravitational force are then considered completely nullified. This is a considered mathematical certainty of the linearly distributed density within a mass as its volume is mitigated by its radius for measure.

However, such a linear measurement would be incorrect if the density of the spherical volume was distributed unequally. Instead of considering just the overall surface area of a mass volume, what if this overall mass volume was delineated as the succeeding layers of apposed surface volumes? And if the densities of these radially decreasing spherical surface areas were defined as succeedingly denser within a discrete body of mass, then actually the opposite affect would be measured. In example, if the density of the mass actually increased toward the center of its mass, in a more exponential representation of its distribution, then the same equation would demonstrate rather an ever increasing affect of the mass' gravitational force. From this latter representation, the further one travels into this spherical volume of exponentially distributed density, the greater the traveler is affected by the gravitational force arrogate of the mass of the whole volume. Exaggerating this example further, where 90% of the mass' density resided in only half the radius size of the original spherical volume, the mass' gravitational force would be increased threefold the deeper one descends towards its center. If this distribution sequence is applied yet again to where 80% of that 90% mass density resided in only a quarter of the radius size of the original spherical volume, then the mass' gravitational force would have been increased more than sixteen times its surface valuation. In effect, when measured to its center, the affects of the mass' gravitational force would be considered amplified rather than nullified.

The expectation is that an increased density toward its center would contribute to an increased gravitational acceleration towards spherical center, without affecting the overall gravitational acceleration of the whole sphere. This increase would then manifest itself in a stratification of layered mass densities within the discrete mass. These layers of stratified mass densities would need to compensate for their increased acceleration with an increased measure of spin. Each such separation in the layered spherical surfaces, or inner stratifications, would be poised to spin faster than its preceding outer layer's surface. It is a combination of forces that would also increase pressure and friction. And this should be expected; for as the inner relationship of these stratifications is realized, it is no more different than the increasing attraction based on increasing mass. In Example: Should stones be piled upon an individual, as the pile increases so does the gravitational attraction of the whole stone pile to the surface of the Earth. Eventually, the increased gravitational attraction of the combined collection of stones would crush the individual. Thereby it can be comparatively viewed that each preceding layer of stratification continues to add to the pile toward the center of the Earth. Similarly, the pressure of diving deeper in the ocean increases toward planetary center. Yet, such amplification in gravitational forces could not be realized upon a simple measure from the original spherical surface area of the whole volume. It is easy to see how one's understanding of the gravitational model would influence their resulting mathematical proofs.

Examining this expectation further, it can also be abrogated that within a sphere of consistent density that the gravitational force is not necessarily constant. Indeed the presence of mass momentum comes into play as well. Subsequently mass momentum is then also required to maintain and demonstrate the overall influence of increased gravitational acceleration towards its center. Consider the gravitational force induced by accelerating in a speeding car; as long as the car is accelerating, the individual feels the related G-Force. And when the car decelerates, the individual feels and reacts to this directional change in the momentum of the induced G-Force. Further, when the moving object is halted instantly, the extreme G-Force is compounded all at once. Add a second moving vehicle from an opposite direction and the gravitational impact can be doubled; e.g., two cars going 40 mph in a head on collision crash will feel the G-Force of crashing at 80 mph. So, too, the gravitational acceleration toward a center of gravity creates a momentum. Similarly gravitational center is influenced by the momentum of gravitational accelerations as well. The expectation is that these opposing gravitational accelerations are continuously colliding head on at planetary center, and other points of opposing direction along the way.

Thus by using the calculations on Earth for example, a constant gravitational acceleration of 0.00983 km/s^2 (or 9.8 m/s^2 or 32 ft/s^2 or 0.0061 mi/s^2 or 127137600 m/h^2 or 417118110 ft/h^2 or 79000 mi/h^2) from opposing directions will collide. The probability of achieving the combined accelerative impact (G-force) of 79000 mi/h^2 upon gravitational center is unencumbered by the opposing gravitational forces that are continuously colliding over the entire diameter of its spherical mass. This is because, as explained earlier, the increasing density is supporting by an ever increasing gravitational acceleration relative to the previously understanding of eight dimensional Space-Time: e.g., s^4/t^4. At first consideration, the acceleration of 79000 mi/h^2 does not represent this impact event in a more generally accepted frame of reference. However, when you apply the concept of escape velocity, the expectation of the collision event becomes more familiar. Using the escape velocity of Earth of approximately 11 km/s (or 36,700 ft/s or 7 mi/s or 25,000 mi/h) as our reference, then, at a minimum, the opposing gravitational forces could collide at approximately 50,000 mph upon reaching gravitational center.

Still as one travels deeper towards gravitational center, the expected escape velocity can be expected to increase proportionately. Subsequently, as demonstrated by the influence of gravitational momentum, the resulting influence of this gravitational collision, relative to its increasing escape velocity, can expressed upon the pressure of its containment. Ergo, the resulting influence of its gravitational momentum beyond gravitational center relates the pressure of these two opposing forces of gravitational acceleration with pressure equivalent to a G-force of 50,000 mph. Such is it that as one travels away from the center of gravity, it relates a weakened influence of the opposing gravitational direction in a lessening of pressure; where upon as one might reach any point on the surface, the opposing gravitational momentum is nullified. In another example, if it were possible to drill a tunnel directly through the diameter of Earth, any organic object dropped into the hole would then gradually experience the increasing pressure of opposing gravitational momentums. Ultimately, the organic object would be squished upon the extreme G-Force of pressure built up between the opposing gravitational momentums. Consequently, the maximum G-Force for any mass exists at gravitational center relative to its relative center of mass density.

The Measure of Matter

Gravity, in my opinion, is better measured via the volume of its influence rather than just by the volume of its discrete mass measurements. For individual mass-to-mass gravitational interaction, this type of considered measurement is sufficient to predict their overall influence upon each other. However, in the grander scope of cosmology, a more comprehensive view of gravity is required to account for the engendering of matter toward measurable mass, even unto cosmogony. In physics, gravity is based on its understanding in the Theory of Relativity as the mass' influence within, or the mass' affect on, the Space-Time continuum. Mass is considered as the principle component of measureable influence for the gravitational force. In this considered involvement of mass, at rest, the density of its volume provides for its unique definability within and influence on the Space-Time continuum. The gravitational constant is a derivative of the force of attraction involved over a surface area in mass of specific density. This force of attraction is then defined as the measure of its gravitational acceleration; an inward acceleration toward a given surface area in mass of specific density relative to its dimensional frame of reference in Space and Time.

Alternatively, this gravitational constant may merely be a consideration of its overall influence as defined by the following formula for mass to mass relations. Where the volume of one sphere is $(4/3) \times \text{pi} \times (\text{radius}^3)$, then the volume of two identical spheres would be $2 \times ((4/3) \times \text{pi} \times (\text{radius}^3))$. An overall spherical volume which includes these 2 identical spheres would then be $(4/3) \times \text{pi} \times ((2 \times \text{radius})^3)$. The relationship of the remaining space [representative of the overall spherical volume minus both of the included spheres] relative to one spherical volume [or half of the overall spherical volume divided by the remaining space] reveals the constant of 0.667 [i.e., $0.667 = (((4/3) \times \text{pi} \times ((2 \times \text{radius})^3)) / (((4/3) \times \text{pi} \times ((2 \times \text{radius})^3)) - (2 \times ((4/3) \times \text{pi} \times (\text{radius}^3)))$

))/ 2] or approximately the value of two thirds. The only thing missing compared to the value of the actual gravitational constant is an angstrom multiplier [i.e., 1×10^{-10}]; the value of 1 angstroms in meters. Using this multiplier of 1 angstrom with the constant value of two thirds and converting to scientific notion results in the value of all mass relationships as 6.67×10^{-11}. As a result of this expression, the relationship of the one sphere volume relative to the remaining space is one third.

Notably an angstrom just happens to be the electrically active distance from the atomic nucleus for electromagnetic and high energy forces (and, as a side note, the gamma ray wavelength boundary within the electromagnetic spectrum). The Swedish physicist, Anders Jonas Angstrom, created this unit of measure to express the wavelength of electromagnetic (EM) radiation in the EM spectrum. Still, while there is a clear association between the behaviors of gravitational and electromagnetic attractions, there has yet to be correlated an empirical equation for this serendipitous association beyond such 'mass shape' relationships.

Alternatively, this electromagnetic constant may merely be a consideration of its overall influence as defined by the following formula for mass to mass relations. The closest approximation of this relationship is half the remaining space as compared to the one spherical volume. If half of the remaining space is two thirds of the remaining volume of the overall spherical volume relative to the one spherical volume, then the one spherical volume is approximately one third of half the overall spherical volume. A similar relationship exists within the one spherical volume, where the resulting relationship for the remaining space [within the one spherical volume] is two thirds of the remaining one third, or 0.22222 repeating. Therefore the non-remaining space of the one spherical volume is 0.11111 repeating relative to the overall spherical volume [as defined in the preceding paragraph]. The only thing missing compared to the value of the actual gravitational constant is an angstrom multiplier [i.e., 1×10^{-10}]; the value of 1 angstroms in meters. Using this multiplier of 1 angstrom with the constant value of two thirds and converting to scientific notion results in the value of all mass relationships as 1.11×10^{-11}. Notably, $1 / (1.11 \times 10^{-11}) = 9 \times 10^{9}$ is the value of the electrical constant for electromagnetic attraction.

Subsequently, the underlying concept of gravitational acceleration is somewhat analogous to the underlying concept of electromagnetic attraction as a representation of the pulling force, or attraction, toward a given surface area. As the pulling force of gravity is mitigated by its distance from the given surface area in mass of specific density, gravitational acceleration is defined as the change in the gravitational force of attraction over time. In fact, both gravity and electromagnetism mitigate this change at a rate consistent with the inverse-square law: i.e., the force of attraction decreases by the square of its distance from a given surface area in mass of specific density. In fact, the energy of a given atomic orbital, upon which all angstrom related constants are based, is also proportional to the inverse square of the principal quantum number.

This reaction, in the case of electromagnetism, is demonstrated via the specific density of its lines of attractive force, i.e., the more dense the lines of force, the greater the electromagnetic attraction. For this electromagnetic attraction, the lines of such magnetic attraction are most dense near the surface area of mass. These lines of force arbitrate a notion of distance relative to given surface area of the volume. Based on this relationship in attraction, whether electromagnetic or gravitational, their acceleration can be considered as a constant because it employs this understanding for change in the force of attraction toward a measured area in volume of mass based on distance. Consequently, the change in distance from the given surface area, in mass of specific density, is used as part of the measure of its volume. In this recalibration of the volume of the mass, the larger volume includes the space or distance beyond the surface area of the mass so as to modify, or decrease, the representation of the mass' specific density. Thus the gravitational acceleration remains constant even as the gravitational force fluctuates beyond the surface area of discrete mass density.

However, gravity and EM remain as distinct concepts in which the understanding of their difference is contributed greatly to the difference of their resultant force; i.e., in terms of particle physics, gravity is about 10^{36} times weaker than EM. Still gravity remains as the dominate force in this most recent evolution of our universe due to the relative difference in densities. While the gravitational force is relative to the density of the atomic mass itself, the electromagnetic force is relative to only the density of electric and magnetic field strengths generated via the electrically charged particles within the atomic mass, such as those that make up plasma (i.e. electrons, protons and other ions). Prior to this most recent evolution in the universe, it is believed that the electromagnetic force reigned supreme. The early universe was defined by its condition in 'maximum Time' and 'minimum Space'. A continuum rich in being without any real atomic mass, potential energy and associated matter formed an electromagnetic universe. It was the medium in which gravitational forces were nurtured with indiscriminate complacency. As the Space-Time continuum evolved from a condition in 'maximum Time' and 'minimum Space' toward a condition in less than 'maximum Time' and more than 'minimum Space', the distinction of matter gave way to the advent of mass densities. Upon the legacy of electromagnetic connections that defined one being in and of the universe, gravitational forces formed among the discrete mass densities of colliding energies. The neural network of universal connections outlined the dimensional boundaries upon which our cosmos was born into the current condition of our evolving Space-Time continuum. An evolution born of its decelerated state of 'maximum Time' and 'minimum Space' was now accelerating toward its termination of 'minimum Time' and 'maximum Space'.

In fact, 'dark matter' was first invented to complement the expected amount of mass in the universe that does not interact with the electromagnetic force. Therefore it is postulated that, for the existing relativistic equations to define the expected gravitational interaction of all matter in the universe, 'dark matter' must account for upwards to 95% of all matter. As an aside, this would mean that the original theoretical model of the 'Big Bang', which was invented to explain the creation and termination of the universe, was based on the observable and measurable effects provided by only 5% of detectable baryonic mass. Still another considered difference is that gravity is defined upon a basis of positive, measurable mass density where upon a measured relationship to the massless particles of 'dark matter' would be negative. Ergo, the concept of a negative gravitational force, or 'dark energy,' can also be related upon the ideal of anti-gravity, or gravitational repulsion. Thus, if the basis for an overall gravitational influence could be recalibrated upon its force of attraction or repulsion, similar to EM influence, it might help to resolve the mystery behind the value of the gravitational constant. The immediate problem with this comparison is that EM is based on the polarity of charge upon a mass of positive density; e.g., a positive mass charge or a negative mass charge.

To aid in such recalibration, theorists had to expound upon the notion of 'dark energy' to explain the anomaly of an accelerating universe in their cosmological equations. Although 'dark energy' was actually first invented to explain the recent observations that the universe appears to be expanding at an accelerating rate, it can also be used to explain the proposed effects of a negative gravitational influence. Even as the exact nature of 'dark energy' is debatable, its expected characteristics must relate to a strong negative pressure. According to the principles of General Relativity, the pressure within matter, or its expressed mass density, contributes to its gravitational attraction on other things. Consequently, should this 'dark energy' be representative of a more negative gravitational force then it could be expected to define a force of gravitational repulsion coinciding with the EM properties of polarity. In other words, mass of positive density attracts mass of positive density and mass of negative density repels mass of negative density.

As empirical science is ill-equipped to measure any form of massless density, it is not surprising that these interactions have eluded detection and that a concept to define these anomalies has eluded discovery. An abstraction of such behavior would subsist upon the recognition of these opposing gravitational forces in complement as representative of their contentious behavior between attraction and repulsion. This recognition could thereby allow theorists to study the interactions of mass with negative density; i.e., 'dark matter' as negative mass density and 'dark energy' as negative gravity. Indeed it is postulated that, for the existing relativistic equations to define the effects of this acceleration in the universe, 'dark energy' must account for upwards to 74% of the total mass-energy. Squeezing this added notion into the current cosmic model, where 'dark matter' would account for 22% of all mass-energy, detectable mass of positive density would only make up 4% of the overall universe. As a further aside, this would mean that the original theoretical model of the 'Big Bang', which was invented to explain the creation and termination of the universe, was based on the observable and measurable effects provided by detectable baryonic mass-energy, or only 4% of the universe.

This is analogous to a clockmaker only understanding the gear to gear ratio in clockworks without knowing how the whole clock mechanism fits together and works to measure time.

Another representative unit of mass with negative density is called the 'graviton.' 'Gravitons' are just as hypothetical as 'dark matter', but they are explained as the quality upon which the speed of light is achieved. 'Gravitons' are thought to be the weave for the fabric of Space-Time such that all existence can be defined upon its measure unit; i.e., the greater the repository of 'gravitons,' the greater its gravitational acceleration. What is notably important about an explanation of the gravitational acceleration in this form is that the notion of acceleration can also be extended to matter in motion; for it is the motion of a 'graviton,' or its energy state, that defines the nature of its mass density. By understanding that the motion of the 'graviton' contributes to the overall force of gravitational acceleration, this motion can be viewed as being assimilated within mass as an increase or decrease in its density. When taken in a perspective, these 'gravitons,' which allow for mass in motion to be collapsed into a model of mass at rest, a 'graviton' can be viewed as the assimilation of matter for mass.

Thereby the motion of the 'graviton' can be calculated upon an increase in the mass' density via its overall inward acceleration toward a measure of its change in Space and Time, i.e., a forced distortion of Space in Time. Herein the measure of this distortion of Space-Time, in perspective of its inward acceleration via the 'graviton,' can be interpreted as increasing the gravitational force of its specific mass density. Similarly the motion of the 'graviton' can be calculated upon a decrease in the mass' density via its overall outward acceleration toward a measure of its change in Space and Time, i.e., also a forced distortion of Space-Time. Herein the measure of this distortion in Space-Time, in perspective of its outward deceleration via the 'graviton,' can be interpreted as decreasing the gravitational force of its specific mass density.

Yet this relationship to specific mass density is still not the complete picture of this forced distortion in Space-Time. This interwoven fabric of Space-Time demonstrates a real measurable influence within the spin of such specific mass density. Graviton distribution, as an integral part of this Space-Time fabric, is relative to the spin motion within this notion of a more complete gravitational influence. Liken to layers of sheer material outside the surface of specific mass density (as substitute for the real effects of gravitational influence beyond the sphere surface of specific mass density), it can be demonstrated that the gravitational force is being dragged along with the spin of specific mass density. This implies that the forced distortion of Space-Time interacts with mass in a way that is more reflective of the pre-proposed model of "immersive liquid inversion" as presented in the first volume (in chapter 4) of this compendium. The relationships of gravitational force to this "immersive liquid inversion" will be discussed in greater detail in chapters 7 and 8 of this book.

Indeed mass, in and of itself, can then be viewed as a forced distortion in Space-Time via its own internal energy state. With some manipulation, gravity would be better represented in terms of the positive density or negative density of its overall volume, however inconsistent, which uniquely defines its influence within Space and Time. This being true, the entire expanse of a predetermined volume would then define, or cause to define, a center of gravitational force in an expression of its introspective interference within the Space-Time continuum as density. Such a perspective would then invoke a kind of 'chicken and egg' analogy, i.e., which came first. Ponder for a moment, did galactic center form first from its black hole(s), to attract and cause mass to amalgamate around it, or rather was galactic center formed from within the entire expanse of its predetermined volume (i.e., its volume of influence) causing a focus of gravitational force toward an aspect of density in which its black hole(s) engendered?

The Measure of Motion: The Twins Paradox Revisited

I would like to take a step back to review concepts that could explain this force of gravity in terms of its acceleration. The first question that might arise should be, "how is it possible to demonstrate that a mass is accelerating if it is not in visible motion?" Because of our reflexive perspective of distance used to formulate an understanding in a measure of acceleration, one would expect to see or imagine a difference in distance over time. One's egocentric perspective tells them that if we are not traversing some distance, then we are not in motion. However, the reality of matter is that everything is always in motion by virtue of its 'being' within a Space-Time continuum. What is not in motion does not exist. Even fixed distances, with which one uses to gauge whether objects are in relative motion to each other, are actually maintained via their constant motion relative to their fixed positions.

In example: While everything associated with such fixed positions are seemingly at rest, relative to one's ability to move about within such a setting of fixed positions, all these fixed positions are all really just moving at the same rate of constant speed. And while these fixed distances are actually only a condition of this consistent constant motion, as coordinated within the Space-Time continuum, they are inexorably being used for measure via their ability to be traversed. Thus it is not simply distance over time that defines visible motion, but rather it is their departure from a more consistent constant motion with respect to these fixed objects. Were everything in Space-Time to remain moving at the same rate of speed, then there would be no observable motion within the universe. Therefore it is this departure from a overall consistent rate of constant speed that determines all objects' visible motion relative to their velocity, acceleration, or deceleration.

This relationship of consistent constant motion also brings to mind the conceptual illusion of coordinated motion. As motion is relative to the observer, the aspect of motion as observable tends to distract from our relative measure of its motion. There is both the illusion of one's relative motion to another object and the illusion of our mind to compensate for one's relative motion. Of the latter, our minds tend to operate consistent with our coordinated motion. If one is at rest, then moving objects can be compared against one's relative state of non-motion as well as against other considered non-moving objects. While these details of one's reality are an aspect of one's ability to focus, the mind tends to image more of an overall perspective with respect to an individual's Space and Time. This allows an individual to map out his fixed setting (or landscape) by focusing on motion that is equal to their own relative motion, while continually scanning that which has greater or lesser motion. In this respect, fast moving objects are blurred unless focused upon. Rather more or less it is a survival technique; those objects of lesser or equal motion are less of a threat to one's ability to navigate within an individual's Space and Time. In example, this is also why individuals cascading toward a head-on collision tend to capture and process these events in relative slow motion. Since the two objects are close in relative speed, regardless of their closing distance, they are able to focus on each other because of the relativity of their motion.

Additionally, the faster an individual travels relative to their own surroundings, the greater their rhythm of processing input and the slower everything around them appears. When one is in motion, the mind tends to compensate for its own movement by comparing the varying degrees of motion about them relative to their own motion. Moreover the mind is quick to adopt this new constant rate in its motion as the norm, allowing an individual to more easily track faster moving objects relative to one's own position. In example, walking fast while staring at the ground flashing by has created a new rate or rhythm for one's individual ability to process its own motion. Should the individual in this example suddenly stop, they would observe the ground moving away from them at a speed relative to when they were walking. While the ground is not actually moving away from them, the illusion represents their mind's greater rhythm for processing input and greater ability to focus on the relativity of their motion. This illusion demonstrates the changing transitional state of one's mind to process input.

This ability to transition between differing rates of motion is also especially necessary to maintain balance and composure while in motion. An individual's motion is compared against what was considered to be at rest and what is considered to be in motion relative to their own condition of constant motion. As an individual tends to accelerate relative to their surroundings, what was previously considered to be in motion (prior to their accelerated state) now appears to be closer to non-motion relative to the individual's current constant motion. In this new condition, the previously determined obstacles now appear to be moving more slowly. The mind actually compensated for one's relative motion by modifying one's perspective ability to process the motion of other objects more relative to their own motion. In part this compensation is fueled by a chemical adrenaline reaction within the mind.

Subsequently this shift in perspective has caused the adrenaline fueled reaction of the mind to modify its ability to input these changes at a faster, or slower, rate as it clamors to regain and retain a balance within its newly acquired surroundings. It does this through a process of perspective analysis by which the mind all but discounts anything that is considered farther away, allowing for one's ability to navigate more fixed positions (or objects moving more slowly relative to their rate of motion) as mere obstacles. This further allows for clarity of thought and the individual's ability to focus on motion that is equal to or greater than one's own relative motion, while continually scanning for lesser motion. Kind of like 'tunnel' vision in this respect, non-moving objects (or objects moving more slowly) are filtered as unnecessary and blurred unless focused upon. The mind tends to image only what it needs for a more of an overall perspective with respect to its own Space and Time. That which can easily be distinguished to have the appearance of slower or non-motion is considered more relative to one's state of motion and less of an obstacle.

While this seems counter-intuitive of Einstein's Special Theory of Relativity, where an individual traveling at the rate of light speed perceives the Present-Time of others (at relative rest) as faster than their own, it is not. A review of this special relativity as associated with Present-Time dilated events can be thought of in two ways: The traveler's environment inside the accelerated condition and the Earth observer's environment outside the accelerated condition of this light speed traveler. Special relativity explains that constants within any frame of reference remain equal; i.e., light speed is 186,000 miles per second whether traveling within a frame a reference equal to 1/1 millionth of light speed or traveling within a reference frame equal to light speed.

For example, shining a flashlight onto a relatively fixed object within the traveler's frame of reference can be measured to travel from the light source to the object at 186,000 miles per second. Outside the traveler's frame of reference, this shining of light demonstrates a perspective, consistent with the Doppler Effect, in which it is not the speed of light (SOL) that is changing but the motion of light relative to the observer. In this way, the SOL traveler is processing information relative to his/her increased rate of motion, while the non-SOL observer's would be processing information relative to their more fixed rate of constant motion. In actuality, while the timed events differ due to this Present-Time dilation, the ability to perceive these Present-Time events alters with the observer's ability to measure and process information.

In this example, a SOL traveler speeding along at 186,000 miles in one second, to or from Earth, would be perceived by an individual on Earth as being in motion (blue of red shifted respectively) respectively. Another way of thinking about this is that the wavelength of light, to or from Earth, would be compressed (blue shifted) or dilated (red shifted). Alternatively the SOL traveler should observe an individual on Earth as being at rest (blue or red shifted) relative to their own SOL motion to or from Earth. However, the SOL traveler's ability to receive and process information is accelerated relative to their accelerated condition. Consequently, while the SOL traveler perceives his/her frame of reference as normal, or consistent to their own rate of processing information, their internal processing clock has fractionalized (or dilated) the time events observed from outside of their SOL frame of reference. In effect, their gravitational acceleration has increased along with their accelerated condition.

Retrospectively it is not the Present-Time dilation of events within the SOL frame of reference, but rather it is the Present-Time dilation of events outside the SOL frame of reference that have been fractionalized. While accelerating to the SOL has imagined this Present-Time dilation as liken to a cryonic suspension of Present-Time in which events actually move slower, Einstein's Special Theory of Relativity would explain that direction has little to do with the effects of Present-Time dilation. So in the example where our SOL traveler is speeding toward Earth, each nanosecond of light received from outside of the SOL frame of reference might be perceived as a second within the SOL observer's dilated condition of Space-Time, irrespective of the compressed condition outside (its blue shifted motion towards Earth). In short, the length of the nanosecond event from outside the SOL traveler's frame of reference remains relatively equal to one second when received and processed within the SOL travel's Present-Time.

Similarly in an alternative example where our space traveler is speeding away from Earth at near light speed, each nanosecond of light received from outside of the SOL frame of reference might be perceived as taking a second to catch up to the SOL traveler's dilated condition of Space-Time. In short, regardless of Einstein's Special Theory of Relativity, the length of the nanosecond event from outside our space traveler's frame of reference took one second of the near SOL space traveler's Present-Time to be received and processed within the space traveler's near SOL frame of reference. Such is it that the Present-Time light events of an individual on Earth are dilated within the space traveler's Present-Time frame of reference. While Present-Time has stretched out for our space traveler, the space traveler's ability to input information from outside their frame of reference has also stretched.

In physics, there is a thought experiment in which the principles of special relativity are applied to visualize this perceived Present-Time dilation called the 'twins paradox'. In this imagined experiment, a twin makes a journey into space in a SOL rocket and returns home to find he has aged less than his identical twin who stayed on Earth. This result appears puzzling because each twin perceives and measures their Present-Time with the same instruments: an atomic clock. In fact, the result is not a paradox in the true sense, but a failure in the imagined experiment. While this paradox can be resolved within the standard framework of special relativity, the expectation that the twin accelerated to the SOL can co-exist within the same Space-Time reference as non-accelerated twin is a fallacy. There have been many explanations demonstrating the fallacy of this experiment, many based upon there being no paradox because there is no symmetry; the SOL twin traveler should have undergone acceleration and deceleration, thus realigning the Space-Time reference for the twins.

However bearing in mind Einstein's Special Theory of Relativity, where direction of motion is not relative to one's accelerated condition, the length of a nanosecond event from outside the SOL traveler's frame of reference still remains equal to one second when received and processed within the SOL travel's Present-Time. Considering that the individual on Earth can only process events relative to their own motion, the length of a second event from inside the SOL traveler's frame of reference is equal to one nanosecond when received and processed from the Earth observer's Present-Time. So while the SOL traveler's Present-Time is dilated relative to the Earth observer, the motion of the SOL traveler would appear more stroboscopic, capturing brief intervals of motion like a fast moving silent picture. Conversely, the SOL traveler would observe the individual on Earth as moving in slow motion relative to his/her own rate of processing motion. So the expectation of where an individual traveling at the rate of light speed perceives the Present-Time of others (at relative rest) as faster than their own appears to be a fallacy; a fallacy based on considering that the SOL traveler's differing Space-Time frame of reference could be maintained relative to the Earth observer's Present-Time upon deceleration. Just as acceleration slows down Time relative to the Earth observer, deceleration speeds up time relative to the Earth observer.

Interestingly enough, such Present-Time dilation represents a condition in which internal and external mass motion is balanced within the Space-Time continuum; i.e., or the conservation of energy via its motion relative to its Space-Time continuum. This suggests that the internal motion of the mass is decelerated (e.g., subject to an induced polarization) when the mass is externally accelerated. This induced polarization is the result of its increasing inertia due to the deceleration of internal mass motion, or the relative increasing (expanding or dilating) of Present-Time toward internal deceleration. In an example relative to the existing Space-Time continuum (outside such an SOL frame of reference) the measure of Present-Time (inside such an SOL frame of reference) is dilated towards increasing inertia; i.e., Time slows relative to increasing inertia.

While at first this appears to express a condition of slowing down one's Time relative to one's Space, or slowing down the aging process, one need only remember that the internal mass motion of the human body can only be slowed so much before it ceases to function at all; i.e., kind of like undergoing cryonic suspension. Therefore what I believe Einstein is trying to describe, via Present-Time dilation, is that Present-Time events differ relative to spacial resistance; i.e., increased acceleration toward minimum spacial resistance results in the relativity for a greater degree of distance (or space) among a lesser degree of Present-Time events.

Such spacial resistance is similarly considering upon the Hafele–Keating experiment, Joseph Hafele and Richard Keating took four cesium-beam atomic clocks aboard commercial airliners and flew twice around the world, first eastward, then westward, and compared the clocks against those of the United States Naval Observatory to test of the theory of special relativity. In a frame of reference in which the clock is not at rest, the clock runs slower, and the effect is proportional to the square of the velocity. In a frame of reference at rest with respect to the center of the earth, the clock aboard the plane moving eastward, in the direction of the Earth's rotation, has a greater velocity (resulting in a relative time loss) than a clock that remains on the ground, while the clock aboard the plane moving westward, against the Earth's rotation, has a lower velocity than the one on the ground, resulting in a relative time gain. It would be interesting to re-run the experiment consistent with constants of acceleration and deceleration in support of the 'twins paradox' fallacy.

Spacial resistance presents another much more ominous aspect for the SOL traveler; imbibitional forces of gravitational acceleration. As a body accelerates, it increases in mass. While this has been presented as expanding Space relative to Time, or Time dilation, it need not translate to an increase in mass volume for the same base density. Instead, what has been observed is that this increase in mass is more relative to increasing inertia via greater mass density from within a lesser shrinking of volume. This means that as any mass object is accelerated toward the SOL, it will disproportionately increase in density with respect to its mass volume. In this way the density of mass can be understood as relative to its inertial frame of reference in Space-Time; i.e., mass density increases with increasing inertia. In other words, mass density increases with SOL acceleration. In example, for a clock of geosynchronous orbit about the Earth and an equal clock on Earth, the time of the clock on Earth is a fraction slower than its equal clock in outer space. This analogy has been over used in Einstein's Special Theory of Relativity to describe twins aging at differing rates relative to their respective rates of acceleration, but without integrating these concepts of inertial changes in mass. Indeed the further one descends into the center of Earth, the greater this time rate differential relative to its degree of gravitational influence.

Motion in Measure of Four Dimensional Time

Having once again strayed from the original topic of this section with such side discussions, I would like to return to the concept of measure in motion. Continuing with the perception of measure in motion, not everyone's ability to deal with this transitional change in perspective is equal. The inability of an individual to adopt such changes in their rate of processing often results in throwing these individuals out of balance. Subsequently this causes their mind to modify its ability to input processing changes at an even faster rate that its own, as it clamors to retain a balance within its surroundings. In point of fact, this loss of balance is the considered cause of motion sickness. This can be complicated by one's perspective even further by flooding our mental processors with a greater number of input variables, like changing direction.

The earlier examples explored the adopting of a perspective rate for motion to or from a fixed position, where smaller objects in the distance become larger or where larger objects are getting smaller (as in flipping around to observe backward motion from behind) respectively. However jumping between such varieties in these perspectives can jolt one's processing rhythm quite drastically. Similarly, looking to one's side would force the mind to process both forward and backward motion at the same time. This sideways perspective can be further complicated by changing direction, like on a merry-go-round. Therefore the best way to combat the causes of motion sickness is to regain one's balance by focusing upon a single direction of motion or a more singular rate of constant motion.

Still, there are other forms of acceleration that can not be observed upon any mental perspective. One such form of acceleration can be abstracted as the potential for motion. The potential for motion can also be related as the attractive or repulsive forces of a mass via its physical presence and composition. In this example, it is a mass' potential to impart a departure from constant motion that can be expressly measured as its ability to be accelerated. Or to put it in another way, it is a mass' force of attraction as proportional to the amount of inertial Space-Time that it has displaced, by virtue of its relative spherical surface density. To understand this displacement effect, one must agree upon a measure of relative consistency in Space-Time without mass. As the predominant condition of the universe is a combination of dark matter and dark energy (i.e., massless matter and energy), then the existence of mass is more an intrusion upon this norm. Such is it that this displacement effect is expressed as if the mass has intruded upon its relative consistent, or inertial, condition of the Space-Time continuum.

In example, as a measure of varied mass energy states, existence can be measured to demonstrate a rippling instability in this relative consistent condition of the Space-Time continuum. In this example, matter is bunched about the nucleus of an atom, likened to a frequency wave, wherein mass energy seemingly transfers along from one apex ripple to another. This departure from the inertial relativity of the Space-Time continuum, or the overall consistent velocity of the Space-Time continuum, is remonstrated upon its potential for acceleration, or its attractive force. This is respectively consistent with the rippled state of matter present in its stratified layers of increasing mass density; i.e., there is a shrinking Space relative to an expanding Time for increasing density. Thus, a mass' potential for acceleration is consistent with the change in its inertial frame of reference from the consistent velocity of the overall Space-Time continuum.

Again, a mass' potential for attraction, or gravitational acceleration, is converted upon a measure of the relative density of its overall spherical surface. For this measure, it is not the distribution of density within a mass, but rather it is the overall density to which a mass relates the potential force of its attraction unto its overall spherical surface. Therefore, depending on whether one chooses either Time or Space as its basis for constancy, either Time is stretched or Space is shrunk upon the intrusion of a mass' overall spherical surface density. That is to say, from the perspective of a constancy based on one's relative Time, Space is bunched up liken to pushing on the edge of some sheer material. Relevant to this example, this material is bunched up like an accordion, but more at the point of its intrusion and less as it extends out beyond this forced distortion. Similarly, Space-Time is bunched up upon the intrusive mass density of its spherical surface, and even more towards its denser gravitational center, but then less as it extends out and away from the surface spherical of its mass density. Subsequently the more that Space is bunched over a constancy of Time, the greater the distance per time is accumulated. From the perspective of the mass, this represents a greater attractive force or increased force of acceleration towards the mass. Such is it that the Space-Time continuum is bunched up about the mass, causing a warp in the rate of its spacing, or the Time of its Space.

Since Space becomes denser within the relative constant of a Time based continuum, Space can be viewed as decreasing relative to Time. Alternatively, from a perspective of a relative constant within a Space based continuum, Time becomes less dense. Thus, Time is relatively stretched or dilated toward the center of the spherical mass density. Consequently, the lessening rate of Time for a given Space, the greater the distance per time is accumulated, representing a greater attractive force of the mass or an increased force of acceleration towards the mass. In this latter example, the rate of Time actually slows down as it approaches the center of a spherical mass density. This is consistent in respects to the defined behavior of Time upon encountering a black hole in Space; the greater the perceived mass density, the more Time slows down and the greater the relative potential for attraction, or gravitational acceleration. Again, a mass' potential for acceleration is consistent with the change in its inertial frame of reference from the consistent velocity of the overall Space-Time continuum. Therefore, density aside, a black hole in Space could also be simply representative of the change in its inertial frame of reference from the consistent velocity of the overall Space-Time continuum.

Readdressing gravitational acceleration from the perspective of a Time-based constancy, consider for a moment how one might better visualize this rippled bunching within the relative constancy of the overall Space-Time continuum. This rippled bunching within the continuum might be better represented as evenly spaced concentric spheres of increasing amplitude toward the intruding mass density. This model would then envision a gravitational acceleration toward the mass as an increase in its relative distance over a constant of time. As a gravitational wave, it could be represented as concentric spheres of increasingly tighter spacing to demonstrate the density of Space within the constancy of relative Time. Whereas Time alters from maximum to minimum, it is its relativity with Space that annunciates this condition of gravitation. Thereby the attractive force of gravitation is more a condition of its relativity within Space-Time; i.e., its Space-Time displacement upon greater density. From this perspective, it is easier to extrapolate upon the change in the gravitational acceleration of a mass beyond the relative spherical surface of its overall density. Thus the gravitational acceleration within the mass can differ as the density of Space-Time displacement increases. Upon this perspective one can more readily understand that the greater the relative density of displacement, the greater the force, or measure, of its gravitational attraction. It appears that the force of gravitational attraction relates the propensity of the Space-Time continuum to return to its former constant condition.

However, is gravity best described via an active force of mass' attraction? Or would gravity rather be better understood as an active force of repulsion pervading against the intrusion of mass' density into the relative constancy of the overall Space-Time continuum? Of the latter condition the intrusion of mass' density, it is the former consistent condition of the Space-Time continuum that is actively pushing against and compacting upon a more submissive existence in mass relative to its ability to be condensed. Pause for a moment to think about this perspective. Perhaps the Space-Time continuum, as the predominant force in the universe, is really the active force in this scheme of gravity as well, repelling against any and all invading disruptions into its vacuum: massless condition. Still another consideration is that both forces are active. Such is it that the constancy of the Space-Time continuum is really the dichotomy of forces in which there is an increase in Space to actively dispel an order of Time, while an increase in Time has a propensity to actively reorder the randomness of Space through dissipation, or expansion. Liken to a memory state of its former condition, the constancy of Space-Time tries to reorder these anomalous conditions of mass. And yet, it is this consideration in the relative constancy of the Space-Time continuum that beleaguers a real understanding of these forces.

Instead, comprehension reveals itself upon the idea that even the relative constancy of the Space-Time continuum is itself only relative to the varying degrees in which these dimensional conditions are combined. That is to say, the dimensional components of Space and Time are in themselves not absolute constants. Rather both Space and Time are only consistent; they are consistent in that their constancy varies in degrees of their influence from a minimum to a maximum. Consequently, the relative existence of the universal forces is engendered and expressed upon their relativity in the varying degrees of Space-Time. In our considered condition within the universe at present, the overwhelming force appears to be the propensity to maintain a dominant massless existence within the relationship of the Space-Time continuum. However, random anomalous events of differing Space-Time relativities are allowed to spin up and persist. It is these differences that define the persistence of mass within an otherwise massless universe.

Closing Statements – The Dimensionality of Space-Time

Considering the advent of spin as the active representation of mass, it suggests more than the simplistic view of the traditional fourth dimensional Space-Time continuum; i.e., three dimensions of Space and singular presentment of all spacial punctuation in Time. Rather, there are also three dimensional variations of Time, as well, as representative of the calculations for (acceleration or deceleration)/time. Think of it as similarly abstracted of Space: s/t demonstrates a linear constant of one direction in Time, s^2/t^2 engenders a more progressive constant maintaining Time in two directions, and s^3/t^3 enlivens a constant consistency in three directions of Time. Maintaining Time in two directions, by measure, prepossesses the bounding of infinite Time within the curvature of its persistence. Whereas maintaining Time in three directions, by measure, prepossesses a bounding of acceleration or deceleration within the completion of its own measured volume. Still even here within three dimensional Time the expression of gravitational acceleration is not enlivened in an amalgamation for the spacial preponderance required of the creation for mass.

However, furthering this notion of dimensional variations for Time, s^4/t^4 does embrace the notion of Time in four directions. By maintaining Time in four directions, by measure, it prepossesses a bounding of (acceleration or deceleration)/time within a vortex of its own impetus. Subsequently the gravitational force (acceleration or deceleration) is a consequence of eight dimensions: four dimension of Time imbibing a complementary four dimensions of Space. A presentiment for the actual model in four dimensions of Space should be: points for line, lines for plane, planes for volume, and volumes for density (representative of the separation of positive and negative density for mass).

This pragmatic abstraction for four dimensions of Space and four dimensions of Time emulates the theorization of our universe being examined in ten dimensions as supported by Superstring theorists. The only problem with an expectation of ten dimensions is the visualization or abstraction for maintaining Time in five directions and/

or, for that matter, the visualization or abstraction for maintaining Space in five directions. The abstraction of this understanding is expected to be the 'holy grail' of physics; in physics, the fifth dimension is a hypothetical extra dimension beyond the usual three spatial dimensions and one time dimension of Relativity. A theory proposed by Theodor Kaluza and Oskar Klein used the fifth dimension to unify gravity with the electromagnetic force. However by maintaining a perspective of Time as merely the fourth dimensional constituent of the traditional definition of four dimensions as three of Space and one as Time (the singular presentment of all spacial punctuation), I believe a Grand Unified Field Theory will continue to elude the scientific community. From my perspective, fifth dimensional Time provides a step beyond gravitational force and fifth dimensional Space provides a step beyond the spacial aspect of density. Perhaps this tenth dimensional perspective can only be truly abstracted as the commutation of both Space and Time.

In eighth dimensional events (understood as traditional fourth dimensional existence), the dominant Space-Time relationships are warped to allow for matter to form into mass. Herein, Time is internalized within matter as energy, required to maintain a mass' density. Ultimately, matter accumulates Time, as this internalization renders less Time to be available within the relativity of its Space-Time continuum. This continual internalization causes a kind of temporal tension, or the degree to which Time is warped relative to its surrounding Space. Subsequently, matter ages as it accumulates more and more Time relative to its Space. In the end, the surrounding Space becomes devoid of Time causing an imbalance that must be corrected. It is a unique perspective, to be sure, that would require some retraining of the mind. But it is not an implausible approach to understanding the forces of gravity. Yet whatever the approach, mass flows towards an overall confluence of Space in Time; i.e., mass travels towards 'maximum Space' and 'minimum Time', increasing in acceleration.

Chapter 7

"The How of Evolution in Existence"

Prologue

So, remanding an earlier premise of existence, what is not in motion does not exist. Yet if all motion is not equally represented, then how does one measure motion for existence? In example, considering a fixed rate of motion, or constant velocity, as the measured amount of time required to traverse one foot, then the amount of time required to traverse two feet is greater than for one foot. Subsequently a comparison of the two given distance segments traveled, from a measure of their traversal time, demonstrates that it takes more time to traverse more distance. Thus the two given distances of one foot and two feet are deemed as unequal. This is to be expected. However when considering a non-fixed rate of motion, or a varying instances of constant velocity, over the same measured distance of two feet, it can be demonstrated that the time required to traverse two feet while moving at a greater constant velocity could be the same as for traveling one foot at a lower constant velocity. Therefore depending on the rate of speed used to traverse these two unequal distance segments, the end result of their traversal time could be the same. Consequently a measurement for motion must also compensate for these differing rate changes in speed of motion.

Yet if the rate of change in speed was due to the environment rather than the rate of the traveler, then the distance traveled could still appear to be the same. This would create an anomaly that might show no compensation for its measure. In this latter example, if the greater constant velocity motion was a consequence of the environment, as in the form of a moving sidewalk, then both travelers could be moving at the same constant velocity. So while both of the traveler's rates of constant velocity were equal, their rates of speed would still vary based on the combination of their speed and that of their moving sidewalk. Overall such a measurement would then demonstrate a traversal of unequal distances in the same amount of time; for although both travelers have only taken one step, each on their own moving sidewalk, one traveler has actually traversed farther.

Even when each of the two given distance segments traveled and their traversal times are measured from the perspective of either traveler, it would appear that both the unit of distance traveled and the time spent traveling are equal. However upon being viewed via a third objective observer, it can be seen that the actual distances traversed are different. So their distances could be deemed as equal relative to each other, and still be unequal relative to their overall Time and Space. And although the distances traversed are clearly unequal from the perspective of a more objective observer, the travelers have demonstrated the illusion of being equal from their individual perspectives. Even if the travelers could realize the measure of these differing distances upon their journey's end, each must still acknowledge that he had only taken the one step at the same rate of constant velocity.

Taking this latter example one step further, consider the rotating motions of individual planets for our moving sidewalks. The Earth rotates at the relatively constant rate of 465 miles per second, while Mercury rotates at the relatively constant rate of 3 miles per second. From a third party objective observer, one step on Mercury displaces the traveler approximately 3 miles and one step from its Space-Time origin. Similarly observed, one step on Earth displaces the traveler approximately 465 miles and one step from its Space-Time origin. Now consider their orbital speeds about the Sun. The Earth orbits the Sun at about 18 miles per second, while Mercury orbits the Sun at about 30 miles per second. Therefore, from a third party observer on the Sun, one step on Mercury displaces the traveler approximately 33 miles and one step from its origin. Similarly observed, one step on Earth displaces the traveler approximately 483 miles and one step from its origin. Venturing further to consider that our Sun orbits the centre of our Galaxy about once every 200 to 230 million years, at a speed of anywhere from 220 km/s to 250 km/s (or 136.7 to 155.3 mps) depending on your reference, then the actual distance relative to an objective third party observer outside our Galaxy is 136.7 (or 155.3) miles plus the single step distances of either traveler. When considering the actual Space-Time origin and destination relative to an observer at a fixed point outside of our galaxy, the traveler's one step is quite large indeed.

Now considering an even further exaggeration of this moving sidewalk analogy, it can be imagined that the entire galaxy is being swept along at light speed. Subsequently the sidewalk is moving so fast that our stepping traveler leaps many light years away with each single step. This is going to be more difficult for the objective third party observer to follow. So, for measure, the stepping traveler needs to signal his location to the objective observer. Yet under these conditions it can be imagined that the stepping traveler is invisible to the objective observer, because even the stepping traveler's light signal will not be seen for many years by the objective observer. For all intents and purposes, the invisible traveler does not even exist in the objective observer's Present frame of Space-Time reference; relegating our observer to be more of an objective spectator than an observer. Neither the stepping traveler nor the objective spectator can sense the other because neither exists in the same measurable Present-Time frame of reference. And although each may exist at the same time, both the invisible traveler and the objective spectator actually exist in differing dimensions of Space-Time relative to their constants for velocity. Such is the illusion of motion that motion is relative to the individual and the context of its limitations, via its ability to be measured.

The magical factor of any motion is the quality of change to be measured over time. And where time becomes a unit of measure, it begets the illusion of being measured. The alterity of time measured is the illusion provided for representations of Space-Time Transport (i.e., the hyper-jumping from one Space-Time location to another). However, the illusion is not so much whether such Space-Time conditions would exist, but rather the illusion is whether or not one could travel between the two differing dimensions of Space-Time instantaneously. The dimensional conditions of motion are only relative to the dimensional conditions in which that motion exists within the Space-Time continuum. Thus, such trans-dimensional travel is not just a transition through differing dimensional conditions within the Space-Time continuum; it is a complete change of its identity required to allow one to coexist in two differing dimensional conditions. Likened to our considered condition of motion sickness, mass is held in balance via its condition within the Space-Time continuum. Altering mass density between two differing dimensional conditions scrambles the balance of the mass from its original form of existence, precluding the ability of a mass to travel instantaneously between two differing Space-Time continuums. Yet this illusion of trans-dimensional travel is but an aside to the original discussion of measured motion.

To continue with this discussion of measured motion, one must review the concept of velocity more closely as a measure of relative motion. The concept of velocity should be understood as that measured unit of a fixed rate of speed as defined by its resultant adjacent measured units of equal distances traversed during given periods of equal time, from one measured unit to the next. So long as each period of time and each measured unit of distance traversed remain equal, then its corroboration demonstrates a fixed rate of speed that remains constant. In compliment, using this corroborated measurement as a constant for this fixed rate of speed over any given period of time being traversed, if the distance is the same then the velocity represented for each distance is the same. Therefore in any equal distance per any equal fixed rate of time, i.e., distance over time (distance/time), the measured rate of fixed speed defines the constancy of its velocity.

Still, each corroborated measure of constancy is rather just another segment of the distinct movements that are relatively equal. Therefore each constant is but part of an overall, larger concept of interacting rates of speeds. When considered in concert, each relative fixed rate of speed can interact to alter the constancy of the overall combined velocities. And yet this change to the constancy of a velocity may also be representative of a fixed rate measure via its changing motion as either acceleration or deceleration. These measures of non-constant velocity, or changing velocity, add another dimension to the measure of a distance traversed. Therefore motion can then also be understood as that measured unit of segmented non-fixed rates of speed, or as a changing rate in speed, as defined by its resultant adjacent measured units of equal distances traversed during differing periods of unequal times for that measured unit. As presented in the previous sentence, the measuring of motion gets quite complex very quickly when all factors are considered in their proper perspective.

In acceleration, where adjacent measured units of equal distances are considered, the time to traverse equal distances decreases. In compliment, for any given period of adjacent measured units of equal time, the distance being traversed increases. Such is it that the fixed rate of speed for the succeeding unit is greater than the fixed rate of speed for the preceding unit. When this occurs, the concept of a velocity in transition can be better understood as that measured unit of distance traversed in a period of two dimensional Time, from one measured unit to the next. Thus, for each measured unit, the fixed rate of speed can be representative of a fixed distance. In this manner, where velocity is measured as distance/time, velocity in transition can be used to express the concept of distance/time per time, or distance/time/time (distance over time^2). By virtue of its aspection during this rate of change in transition, the resultant interaction can be defined as an acceleration or as a deceleration (acceleration | deceleration == velocity/time). This then provides a new notion of constancy as well; So long as, for each measured unit of equal distance traversed, the given periods of two dimensional Time remain equal, then the rate of change for each measured unit in two dimensional Time remains constant.

As previously explored, the concept of a non-constant velocity is representative of a measure in acceleration or deceleration. However, the concept of a non-constant acceleration, or non-constant deceleration, does not have any specific terminology, and is only represented in terms of related mathematical formulas. This notion of a non-constant acceleration, or a non-constant deceleration, adds yet another dimension to this measure of distance traversed. Similarly the measured unit of a non-fixed rate of acceleration, or deceleration, is defined by its resultant adjacent measured units of equal distances traversed during a given period of unequal two dimensional Times from one measured unit to the next; i.e., velocity/time per time as either acceleration over time (acceleration/time) or deceleration over time (deceleration/time).

So where velocity/time was noted as a measured unit of distance traversed in a given period of two dimensional Time, acceleration/time or deceleration/time would be considered a measured unit of distance traversed in a given period of three dimensional Time (or distance over time^3). And yet from the perspective of a fixed volume of unique time, it might be stated that acceleration/time or deceleration/time can also be considered as the measured unit of three dimensional Space within a period of three dimensional Time (or Space cubed/Time cubed), for six dimensional reference in Space-Time. The interrelationship of these Space-Time transitions are such that constant velocity (or distance over time) will mimic the unending straight line of one dimensional Space, velocity/time (or distance over time^2) will be expressed as the never closing curve of two dimensional Space, and velocity/time^2 (or distance over time^3) will emulate the full circle closure of three dimensional Space.

The Abstracting of Motion for Existence

These concepts can be abstracted upon the unique perspective of their dimensional interaction; for it is only by their dimensional interaction that an infusion of relativity can be defined. Such is it that relativity is bounded by the notions of what exists and what does not exist. These bounded notions of dimensional interaction for the infusion of relativity can be modeled in eight dimensions (four dimensions of Space and four dimensions of Time). However since relativity involves more than eight dimensions, these bounded notions are better abstracted among two models for ease of its comprehension. Bear in mind that when assimilating this Space-Time representation using two models, as delineated in figures 7.1a and 7.1b below, that the concepts of Space, Time, and Velocity are relative to their measured degree of dimensional interaction. Hence the need to provide a depiction of this interrelationship is better idealized in more than eight dimensions via these models. Yet in relating these cubical dimensionings for Time (T), Space(S), and Velocity (V) in eight dimensions, it is also necessary to provide a presentment of an apriori abstraction from within non-existence as the calculated representation of modeling towards a measure of existence. Therefore, in relating a presentment of this apriori abstraction, the cubical dimensionings of No Time (No T), No Space (No S), and No Velocity (No Velocity) are employed toward a measure in existence. Herein these two modeled abstractions the measured degree of their dimensional interaction actually reevaluates their relationships within Space and Time in an expression of their preternatural measures toward infinity; i.e., Infinite Time (ooT), Infinite Space (ooS), and Infinite Velocity (ooV). Such is it that by bounding the notion of a sixth dimensional measure for acceleration, or deceleration, its presentment conceptualizes an eighth dimensional interaction by virtue of its expressed condition of the relativity in Space and Time; wherein infinity becomes the bridge between Existence (E) and Non-Existence (NE).

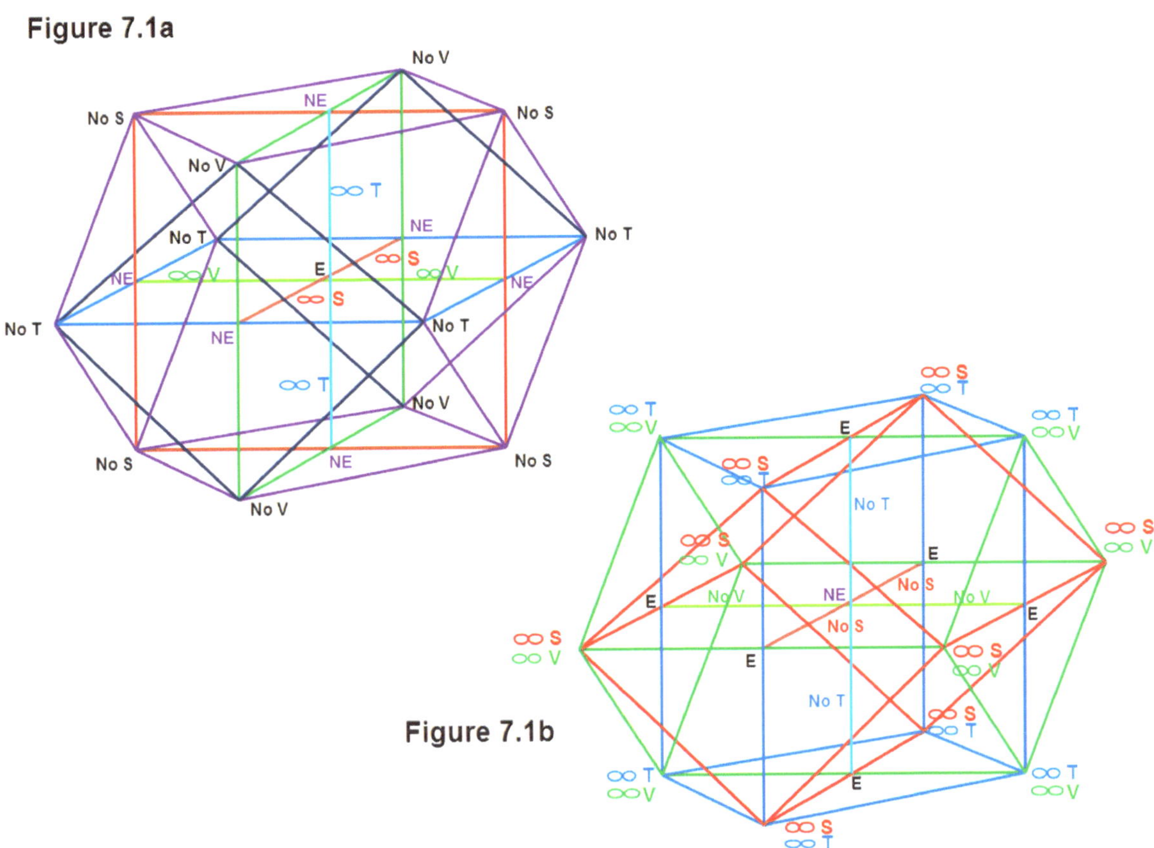

Figure 7.1a

Figure 7.1b

Previously, it was noted that time decreased upon a measured unit of distance for greater speed; or the same measured unit of distance was traversed in half the time. Another way of representing this concept, instead of decreasing time per unit of distance, is to stretch this same unit of time over a greater distance. This is similar to judging one rate of speed greater than another by noting the distance traversed over a set segment of time; i.e., 10 miles per hour as compared to 100 miles per hour. So, in effect, we have stuffed more distance into a given period of time. Using the measured unit of a three dimensional distance, it has the resultant effect of stuffing distance into a fixed period of unique time that expresses dimensionally greater velocity. Extending this understanding of a three dimensional distance towards a measured unit in fourth dimensional distance, extrapolates the concept of an increasing velocity over time.

Further extending this understanding of a fourth dimensional distance towards the unique measured unit of a fourth dimensional existence, extrapolates the concept of an increasing acceleration over time. It is an extrapolation that bounds infinity in the internalized measure of increasing acceleration; an internalization which preordinates a focus of force for a presentient form of matter. It would be at this point that its linear measure of distance is no longer to be perceived and its motion upon distance becomes insignificant to the calculation.

Its relative 'being' is no longer expressed as motion in one direction. Rather its measure is representative of its motion in all directions. As its motion in measure is internalized towards the presentiment of matter, Time is being internalized in Space. Another way of putting it might be that its relatively bounded being presents a minimizing of Time in Space, or the maximizing of Space in Time, upon its increasing acceleration toward an eighth dimensional persistency. It is actually expanding the Space of its relative being by internalizing the Time of its relativity. Moreover, it expresses a force of potential matter via its measured expansion in eight dimensions; i.e., massless matter, or dark matter if you will. In doing so, it has divided 'being' in the universe into a force of potential matter and that which is not.

Subsequently, we have answered the first question. Pursuing this vision of in a visual abstraction, it would be better to switch our train of thought to use Space and Time rather than distance and time. Herein a fourth dimensional condition, one can envision that Time has been stretch over a very large portion of Space or, it can be said, that a very large portion of Space has been stuffed into a unique unit of Time. In review, one needs to remember that maximizing Space is akin to a measure of greater distance and less resistance; e.g., the absence of mass. The resultant potential matter, which is the subject of an increasing acceleration over time, is now measured to include the density of its potential force, or the non-density of matter. Another way to envision this perspective can be expressed as the amount of Space stuffed into a unit of Time as a presentment of potential matter. Herein the definition of density lies in its force of being, or the internalization of Time, and not in its mass. Not only can one imagine the notion of an increasing acceleration in massless matter, but one can also envision how its non-constant acceleration relates to the density of its measurement. Where there is greater acceleration over time, there is greater density of force relative to its surroundings. This is the precept of the force that defines gravitational acceleration. Indeed, herein one is able to view cosmologic black holes as the ultimate result of their increasing acceleration. Wherein mass is the decelerated condition of massless matter, mass retains an affinity to return back into this resilience condition. Ergo, a black hole represents the balance between mass and massless for both matter and energy. A galactic black hole is the focus of dark energy upon the accumulated mass of the galaxy, or dark matter. Thus, considered travel into a black hole represents dissolution into massless density.

The Universal Balance

The second question which should arise is why does this acceleration manifest itself as an inward attracting force? It has been demonstrated that massless matter is really a variant product of its own acceleration over time and that the acceleration is based in the abstraction of mass density, i.e., its stuffing of Space in Time via measure. Why then does this variant product of increasing acceleration express itself as a force which exhibits an inward attraction rather than an outward repulsion? Immediately it might be explained as the primordial condition of its dimensional condition, but it is more than this. I believe there is a universal balance to be considered. Just as heat and cold must be equalized, so it must be true of such forces as density and expansion. Even further, it must also be true of existence and nonexistence. For every type of distortion in the Space-Time continuum, such as where the density warps the medium of Space in Time, there is a complementary distortion which can be imagined within the medium of Space and Time. Such an antithesis might be envisioned as the density of Time in Space. Such is it that the absence of spacial density can also be viewed as the complementary warp within the medium of Space and Time. In this way it appears the overall universal balanced is maintained within smaller degrees of Space-Time distortion by maintaining still smaller pockets of these localized conditions; i.e., greater mass versus greater potential.

In the case of mass, for example, the absence of mass is being squeezed out from the engendered density of mass, much like squeezing a water soaked sponge. Where water represents dark matter, the more the sponge is squeezed, the more water is force out and the denser the condition of the sponge. Yet if we relax the squeezing of the sponge, allowing the sponge to expand upon its potential resilience to regain its former shape, then the water quickly rushes back into the sponge and the balance is maintain. So too, can the absence of mass allow the potential of matter to expand. As stated of Einstein's Theory of Relativity, mass expands with acceleration. Yet the increase in mass is not relative to its size. Rather the increase in mass is relative to its increased density. This is somewhat of an oxymoron when considering that a mass becomes denser as it expands it measured influence. Yet relativistically, it relates to the balance of its measure rate of influence due to its increased density: a measure of its dimensional subsistence. In fact, dark matter is actually being absorbed as the potential of the mass is being expanded. Or another way of putting it is that the dark energy is applying force of increased pressure on mass at some sub-component level.

Subsequently, these sub-components of mass are becoming denser while expanding away from each other, resulting in an overall increase in its mass. This inward repulsive force of dark energy against kinetic mass is relativistically related as a force of attraction accredited to the subsistence of the mass. However, what is important to note is that mass persists as a force of attraction because it is expanding against dark matter by virtue of its mass acceleration. Thus mass acceleration is balanced via the gravitational acceleration of the surrounding medium of Space and Time due to its increasing potential. Another way to approach such an abstraction is to state that the density of mass is balanced via the absence of its emptiness, or massless matter, within its Space-Time continuum. The more dense a unit of mass is by this measure, the less its emptiness and the greater the force of attraction to it; i.e., the greater its gravitational acceleration. Subsequently, this answers the second question.

The third question, which should arise, is why doesn't everything manifest itself as an inward force of attraction? A corollary to the density of mass, manifesting an inward force of gravitational attraction, is that very small objects have little or insignificant mass. The lesser the discrete mass, the less significant the measure of the discrete mass density and subsequently the less significant the measure of its manifested force of gravity via its gravitational acceleration. Yet upon this corollary, it can also be seen that the greater its non-density relative to its surrounding medium, the greater its outward force of repulsion. So there is a consistent balance that defines the subsistence of matter and being in the universe.

As presented in volume one of this book series, existence is as defined by its measure of Time and Space. Time and Space express measure via their direction towards a maximum and minimum. If the squeezing of Space into a subsistence of discrete mass density manifests an inward force of attraction, then the expanding of Space into an insistence of discrete absence of mass density should manifest an outward force of repulsion via its force of deceleration. In maintaining the balance of the universe via such a measure in Space and Time requires that there must be a way to represent this measure unit of non-mass, i.e., there must be a way that Time can be stuffed into Space. Complementary reverse engineering would propose that the outward force of repulsion must be the result of its insistence of identity interacting within its medium of Space and Time. This is evidentiary of degenerative pressure, a term developed to explain Pauli's exclusion principle, wherein a repulsive force is the reactive result of greater external density. Expressing such an abstraction of this outward force remands a complementary perspective of the previously discussed fourth dimensional distance as the variant product of increasing deceleration over time (see figures 7.2).

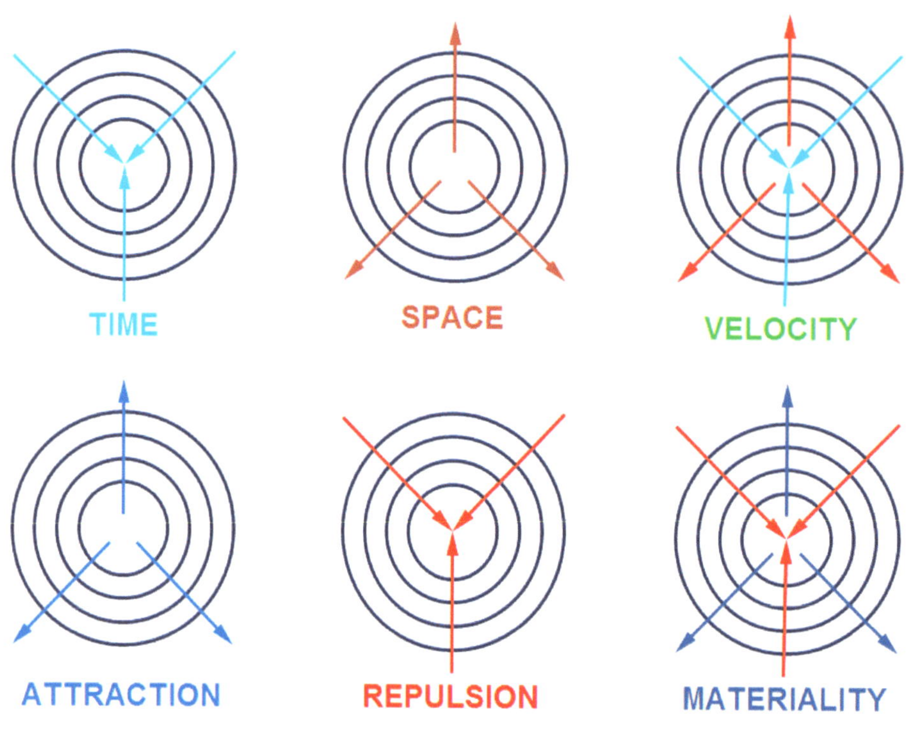

Figures 7.2

To re-approach this complementary perspective, it would be best to re-review the concept of velocity as it relates to mass in motion. The concept of velocity should be understood as the measured unit of a rate of speed as defined by its resultant measured unit of distance traversed from one given period of time to the next. So long as each period of time for the measure unit of distance traversed remains equal, the rate remains constant and therefore, by definition, the velocity is constant. Since, for any given period of time, the distance is the same, velocity is usually represented by its speed, i.e., distance per time or distance over time (distance/time). Subsequently the concept of acceleration adds yet another dimensional perspective to this measure in that it should be understood as the measured unit of an increasing rate of speed as defined by its resultant measured unit of distance traversed from one given period of time to the next. In acceleration, the given period of time decreases for the measured unit of distance such that the speed of the next period is greater than that of the speed of the first given period. When this occurs, the velocity itself becomes the measured unit of a two dimensional distance traversed from one given period of time to the next, i.e., distance/time per time or velocity over time (velocity/time). This provides a new notion of constancy as well; So long as each period of time for the measured unit of a two dimensional distance traversed remains equal, then the rate of change remains constant and therefore, by definition, the acceleration is constant. As velocity is a measured unit of a two dimensional distance, acceleration then became the measured unit of a three dimensional distance.

However the concept of deceleration adds yet another variation of this dimensional perspective to this measure. Deceleration should be understood as the measured unit of a decreasing rate of speed as defined by its resultant measured unit of distance traversed from one given period of time to the next. In deceleration, the given period of time increases for the measured unit of distance such that the speed of the next period is lesser than that of the speed of the first given period. When this occurs, the velocity itself becomes the measured unit of a two dimensional distance traversed from one given period of time to the next, i.e., distance/time per time or velocity over time (velocity/time). This provides a complementary notion of constancy; So long as each period of time for the measured unit of a two dimensional distance traversed remains equal, then the rate of change remains constant and therefore, by definition, the deceleration is constant. As velocity is a measured unit of a two dimensional distance, deceleration now becomes a measured unit of a three dimensional distance.

Extending this understanding of a three dimensional distance via deceleration toward its measured unit as a fourth dimensional distance, extrapolates the concept that rather externalizes a measure of greater deceleration from within presentiment matter. Thus while its motion in measure is internalized towards the presentiment of matter, as Time is being internalized in Space, Space is being externalized in Time. As Space is externalized in Time, it abrogates a focus of force for a presentient form of energy. It would be at this point that its linear measure is no longer being perceived and its motion upon distance becomes insignificant to the calculation. Its relative 'being' is no longer expressed as motion in one direction. Rather its measure is representative of its motion in all directions. As its motion in measure is externalized within the presentiment of energy, Space is being externalized in Time. Subsequently it then expresses a view of kinetic matter, or energy at rest. Herein its measure of eighth dimensional insistence, one can envision that Space has been abated within a very small portion of Time or, it can be said, that a very small portion of Space has been centripetally dispersed within a unique unit of Time. The resultant energy, which is the subject of this increasing deceleration over time, is now measured to include the absence of Space involved; i.e., the amount of Space centripetally dispersed within a unit of Time. Not only can one imagine the notion of deceleration in the absence of mass in motion, but one can also understand how deceleration relates to measurement of the density of its emptiness. Where there is greater mass deceleration, there is greater absence of mass density. This is the precept of force that inverts a definition of gravitational acceleration towards a manifestation in kinetic energy.

An alternative to the visual modeling of deceleration as a repulsive force is the mental abstraction of the relationships of distance and time via a measured unit of a rate of decreasing velocity. Previously, it was noted that time increased for a measured unit of distance for lesser speed, i.e., the same measured unit of distance was traversed in twice the time. Another way of representing this concept, instead of increasing distance per unit of time, is to compress the unit of time for a set unit of distance. This is similar to judging one rate of speed lesser than another by noting a set distance traversed over a decreasing unit of time, i.e., 100 miles per minute as compared to 100 miles per hour. So, in effect, we have stuffed more time into a given unit of distance. Using the measured unit of a three dimensional distance, the effect of stuffing time is exponentially greater. This understanding of dimensional distance could then be extended to a concept of a increasing deceleration such that it would employ the understanding of a measured unit of a four dimensional distance. It would be at this point that the mass would no longer be perceived to be in motion as time becomes insignificant to the calculation. Rather it would express a view of kinetic energy (i.e., non-mass animism). Pursuing a vision of such a visual abstraction, it would be better to switch our train of thought to use Time and Space rather than distance and time. Herein a four dimensional distance, one can envision that Space has been stretch over a very large portion of Time or, it can be said, that a very large portion of Time has been compressed into a set unit of Space. The resultant energy, which is the subject of this increasing deceleration, is now redefined to include the density of Time involved; i.e., the amount of Time compressed for a unit of Space. Not only can one imagine the notion of an increasing deceleration in energy at rest, but one can also understand the how increased deceleration relates to the non-density of its measurement. Where there is greater deceleration over time, there is greater absence of Space.

Yet, where our study of these manifested forces began with a density greater than its medium of Space and Time, following a concept of density less than this medium of Time and Space does not subject these forces to termination at the equalization of its density to the medium in Space and Time. Wherein mass deceleration abates the densities achieved via of mass acceleration, its momentum abrogates its former relationship within the Space-Time continuum. Rather mass deceleration will continue beyond its absence of mass density to become an inversed measure of mass being, maintained somewhat like a bubble. In the absence of its mass density, the density of Space is being absorbed from the engendered absence of Space, or density of Time, analogous to the operation of a vacuum cleaner using a balloon filter. The more the mass is drawn out to and through its balloon filter into the surrounding medium, the more dense the mass build up of the surrounding medium and the less dense the condition of the medium inside the balloon. Where greater mass density is forced unto the surrounding medium, the growing gravitational acceleration of the balloon surface is forcing the containment of lesser mass density to expand. The overall concentration of greater mass is contained outside this expanding balloon of non-density. Recalling the previous relationship of gravity via acceleration, the more dense a unit of mass by measure, the less its emptiness and the greater the force of attraction to it. From an alternative viewpoint, it might be said that mass density is not the result of a force of attraction, but the product of mass absence of density via a force of repulsion. In this case, the relationship of gravity, via deceleration, is maintained via the absence of mass at the concentric focus of its non-density, where the greater attraction is directed outward toward its surface periphery. In the inverse of its previously understood purpose, gravitational attraction is redirected away from a concentric focus as the embodiment of a repulsive force. Imposing itself as a reciprocal of Einstein's energy equation in which mass expands/increases with acceleration, where mass is decelerated then it would be expected to be compressed, or decreased.

This is somewhat of an oxymoron when considering that a mass becomes less dense as it compresses, but this relates to the understanding of a measure unit of rate upon a four dimensional distance: a measure of its fourth dimensional insistence. However, what is important to note is that the massless condition persists as a force of repulsion because it is being compressed by virtue of its mass deceleration. Liken to a bubble, its mass acceleration accumulates on its surface periphery and away from its anti-gravitational center. Decelerated mass is balanced upon its 'gravitational deceleration' within the surrounding medium of Space and Time. Another way to approach this abstraction is to state that the absence of mass density is balanced via the absorption within the medium of the Space-Time continuum. Therefore the less dense a unit of mass is by this measure, the more its emptiness and the greater the force of repulsion that is maintained via its gravitational deceleration. Subsequently, an understanding of this new concept of a gravitational deceleration answers the third question.

Consequently, it is important to note that the two opposing gravitational conditions facilitate the universal balance while fostering its evolutionary development. Herein where gravitational acceleration is bound in Time, gravitational deceleration is bound in Space. For gravitational acceleration, the alternative dimensional boundary is 'minimum Time' and 'maximum Space'. And for gravitational deceleration, the alternative dimensional boundary is 'maximum Time' and 'minimum Space'. Where gravitational acceleration assumes greater density towards its interior, nearest to its center of influence, gravitational deceleration assumes greater density towards its exterior, furthest from its center of influence. Subsequently, gravitational acceleration is manifested via the density of its mass in matter for an inward force of attraction, where gravitational deceleration is manifested via the absence of its mass in matter for an outward force of repulsion. Or to relate it in terms of their resultant conditions, where gravitational acceleration is a product of its kinetic subsistence in mass, gravitational deceleration is a product of its potential insistence in energy. Via this reasoning, the universal being is balanced upon the overall condition of its expansion. As a complement of Einstein's energy equation, the fact that the universe is expanding is relative to its overall condition of increasing acceleration.

It is probably not difficult to understand why these gravitational conditions become easier to manipulate at their cusp, where the differences between them are in constant contention. Using the atomic model as an example of these conditions on the cusp, note that when density (i.e., a degree of gravitational acceleration) is introduced into a sphere of energy (i.e., a degree of gravitational deceleration), it enhances the atom's potential force to attract and dilutes the atom's kinetic force to repel. This causes the atom to contract its volume of influence by losing energy. However, when density is drawn out of this sphere of loosing energy, it enhances the atom's kinetic force of repulsion. This, in turn, causes the atom to gain energy and expand, or extend, its volume of influence. Such is it that this loosing energy transmutes to enhance the periphery of its non-density in the bubble-like containment of its sphere of influence.

This is representative of Neils Bohr's application of quantum theory on Ernest Rutherford's theory of the atomic structure. Ernest Rutherford's model suggested that electrons would contract and expand their distance to center of the atomic structure to account for fluctuations in either losing or gaining energy. In this sense, the electron's position would represent the absence of density reflective of gravitational deceleration and the proton would represent the substance of density reflective of gravitational acceleration. The addition of a nucleus would seem to suggest yet another layer of commingling within the predominant influence of a condition in gravitational deceleration; liken to a spherical representation of wave propagation rippling in degrees of these commingling gravitational conditions. The resultant entropy, which is the subject of a decreasing deceleration, and/or decreasing acceleration, over time, is now a measure in the contention of density within the matter involved (i.e., the amount of Space centripetally parsed within a centrifuge of Time), propagating variant kinetic and potential relationships in matter from which to engender eighth dimensional purpose in energy for mass (see figures 7.3a and 7.3b). This is the precept of matter that parses the gravitational conditions of deceleration and acceleration unto measured unit forces of charge for repulsion and attraction. But I am getting ahead of myself. These relationships are more a subject for my next chapter.

Figure 7.3a

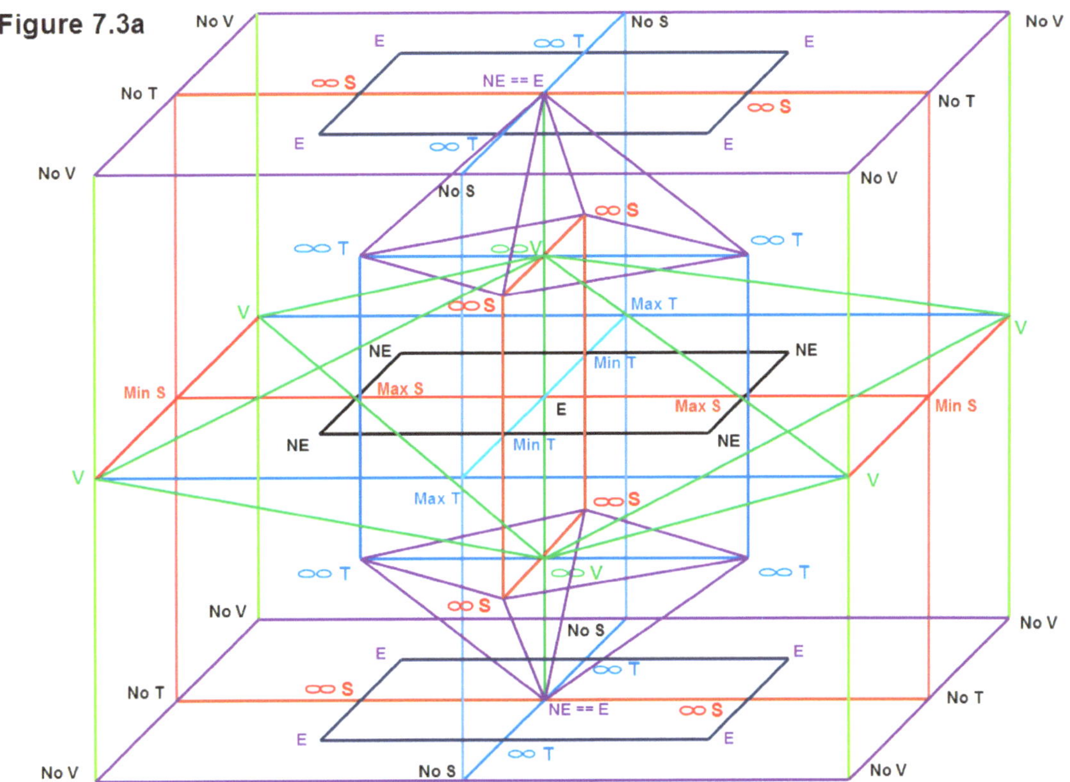

Evolution of Existence toward Potential Conception

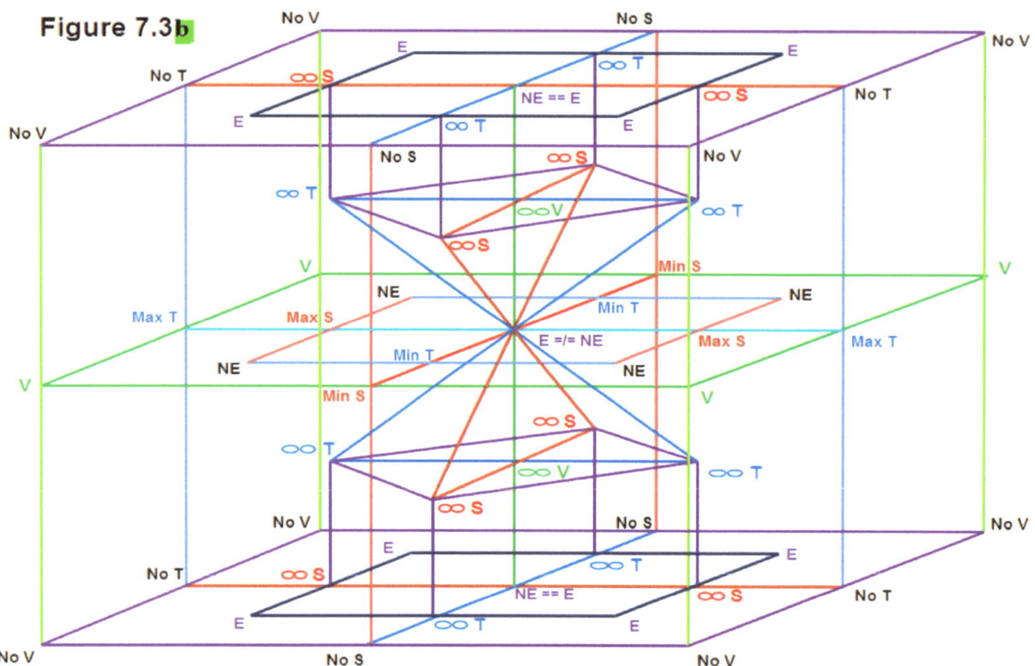

Figure 7.3b

Potential Conception Toward Kinetic Evolution

Closing Statements - In the Shadow of Light

Therefore, via gravitational acceleration, it could be stated that 'Gravity' arises via the potential force of matter to be promulgated from the foci of its predetermined volume. Subsequently, via gravitational deceleration, it could be stated that 'Gravity' arises via the kinetic force of matter from a circuit of its encompassed volume. In the extreme, gravitational deceleration would perhaps be more akin to a notion of 'dark energy' as it relates to a force of matter in the absence of mass; defining the scope of an inverse force of massless density. In other words, there need not be more mass to account for the discrepancies in the quantum calculations of the mass based universe (i.e., based in terms of gravitational acceleration), because there is an active force of gravitational deceleration, "dark energy" if you will, in which the universal equilibrium is being maintained. At the very least, if one wishes to refute this measure of a gravitational deceleration, one would need to recalibrate how one measures minimum mass. However, if one embraces this measure of gravitational deceleration, it builds upon our present understanding of the relativity in a Space-Time continuum. Wherein gravitational acceleration pervades as the glue of existence in an expression of Space bound in Time, and gravitational deceleration imbues existence with the kinetic animism as an expression of Time bound in Space. Together their synergy imports the prodigy of an existence from non-existence in which cosmogony is engendered. And yet this prodigy is but only a portion of the alternative universe model, as presented in volume one of this book collection. And as transmuted within these varying extremes of energy and mass, from a perspective in measure of their gravitational conditions, an evolution of this prodigy is but only part of the enigma that is our existence.

Consider a descriptive abstraction of a black hole within the perspective purview of this alternative universe model. Consistent with the event horizon of a black hole, acceleration/time becomes the energy for mass density; a density formed and nurtured in its own reflection of deceleration/time. Deceleration/time becomes the energy of non-density in the form of a graviton; the greater this negative density, the greater its reflective gravitational acceleration. Remember, the graviton is a theoretical quantum particle that must be massless to support the notion of an unlimited gravitational range within our universe. So as acceleration/time is measured beyond the capacity for light, to remain subsistently autonomous, it is lost in it own shadow. This shadow of light presents the spherical nature of its dimensional transition (e.g., 3 dimensions of Space and 3 dimensions of Time), as photons are polarized into the gradient constituents of their condition: positrons and electrons. As polarization increases, a paralytic permittivity disengenders quark bonding for the distillation of potential force. As this dimensional transition is abetted in the form of its distilled transmutation, it creates fields of gravitational and electroweak energy in the wake of its temporality.

The dual axles of the gravitational and electromagnetic forces become more apparent in the liquidity of their specialization, supportive of their conditions in Space-Time. A condition, similar to the pre-proposed model of "immersive liquid inversion" as presented in the chapter 4 in the first volume of this compendium, emerges in the separation of these forces along their distinguishably individual field axles. In the distillation of transmutated mass, the graviton begets the shadow of light, or the propensity to exist in the absence of mass. In example, this shadow of light conveys the unique perspective of the event horizon for a black hole; where the perception of the horizon is most pronounce in its lost reflection. So whether the black hole is truly spherical or simple an illusion of its event horizon is irrelative to its temporal condition of acceleration. Its density is fueled as much by its condition of accelerating acceleration as it is by its dimensional transition. For even though the gradient constituents become ever less dense, they allow for the compaction of the mass as whole, distilling and dispelling ever more gravitons. The build up of gravitons outside of the compacted mass will eventually define a volume of negative density much greater than its positive density. Liken to a balloon "bubble" which is inflated beyond its breaking point, mass acceleration reaches its point of dimensional reflection for mass deceleration, turning itself inside out to regain balance. Or another way to put it is that it can not continue accelerating beyond the limits of its defined Space-Time continuum (i.e., the confinement of its dimensional condition). Analogous to the over inflated balloon, the mass deceleration mirrors its mass acceleration in an instaneous reversal of Time in Space, trading potential force for kinetic energy. By definition, any change in constant velocity that is not the result of acceleration, must be the result of deceleration. It is this dimensional mirror which results in the rapid decompression of its compacted mass for the recombination and redistribution of its atomic matter.

An interesting fact about the condition of the simple soap "bubble" is that there is a tremendous amount of energy required to maintain its shape. The expectation of this energy support is that it will eventually reach a point of reversal wherein this energy will all be released at once. Although it does not present quite the display of a balloon popping, liken to a supernova, it still releases all of its energy at once in that one instance. The difference in the effect is how the bubble is burst. A soap bubble maintains somewhat of a "happy-medium" within and without its condition. Add another soap bubble and the two can combine to form a larger bubble; Split a soap bubble in half and the two can survive as smaller bubbles. A soap bubble, it would seem, only burst as a result of its loss in viscosity and not much energy is expended.

A balloon, which is being blown up, is actually building up an elastic force which is attempting to squeeze the space inside the balloon to maintain its condition. The expectation of this support for energy is that it will eventually burst with the reversal of its elastic force being expended. The result would be similar if the build up of the elastic force was actually due to the creation of a vacuum outside the surface of the balloon. However, what would be the fate of a balloon in which the vacuum was being created from within the balloon. In this case, there is actually a build up of elastic force which is attempting to expand the space inside the balloon to maintain its condition. This example is similar to the previous example wherein a vacuum was created outside the balloon as the elastic energy of the balloon was used to condense the air and collapse the balloon on itself. However in this example, it is difficult to imagine the balloon stretching or that it could reach the point of reversal and burst, because it was being squeeze into a more dense condition. When would it reach the point of reversal wherein its energy would all be released at once?

Now, what if Time and Space presented a modified continuum whereby the balloon wished to remain in its static condition of being blown up and a vacuum was created inside the balloon? This is the predicament of mass which attains too much energy in the form of gravitational acceleration. It reaches a condition in which it builds up energy to the point of reversal, i.e., a black hole, and then bursts with the expended force of the energy it built up trying to maintain its condition. However, unlike the implosion of a supernova, the black hole presents a modified continuum of Space-Time in which it supported within a graviton shell much like a soap bubble. Its expended energy defines the event horizon of the black hole. And, like a soap bubble, it attempts to maintain a happy medium between its external positive mass density and its internal negative mass density. As a consequence of this design, comingling the attributes of the balloon and bubble, the black hole provides for the expectation of an almost perpetual being. Subsequently the demise of a black hole would be limited to the conditional viscosity of its graviton shell. At what point this conditional viscosity might be interrupted is difficult to abstract as the density of the internal vacuum differs from the external vacuum of the surrounding space. Perhaps, like the soap bubble, a black hole will be snuffed out in the whimper of its dimensional transition. Or perhaps, as some would like to theorize, a black hole demise becomes a white hole (or Big Bang) birth of yet a whole other universe.

In this modified example, it is easier to imagine that the balloon could reach the point of reversal because it has a space to stretch into, and a place to burst into as well. Remember bursting, like any explosive force, directs it force toward least resistance. In this modified example, the direction of least resistance is into the less dense continuum inside the balloon. On a scale of the atomic model, this implosion would compel the intermingling of negative and positive forces of energy to combine by degree into a variety of Space-Time distortions. And yet, until it actually burst, how much might the "balloon-bubble" shrink before reaching the point of reversal. On the scale of the atomic model, I imagine its behavior to perform much like Neils Bohr's application of quantum theory on Ernest Rutherford's theory of the atomic structure, in which it could shrink and swell in concentric distances consistent with set levels/packets of energy. The atomic model itself is just an extension of the nuclear forces required to maintain it condition. Were this "balloon-bubble" of the atomic model, in the modified example, be shrunk enough to breach the nucleus and reach the point of reversal, it would burst with a force consistent with the energy required to maintain the containment of the atomic model in its more stable, rest state.

Investigating the gravitational well of a black hole, from a perspective of the model as represented in the first volume of this compendium, the dichotomy of motion exhumed from its dimensional morphosis reveals the bounded condition of its mass acceleration. Relative to a perceived slowing of Time in Space, its forward motion is suppressed into a revolving equilibrium of its accelerating acceleration, or Time cubed. The nature of the containment of the black hole is representative of Space-Time cubed, or the sixth dimensional state of Space and Time: 3 dimensions of Space and 3 dimensions of Time. This condition of revivification distends the Time of its Space and redefines it motion of acceleration along its revolutionary axis. The only problem with this transmutation is that the mass has been consumed and digested by the black hole to form energy needed to fuel its ability to maintain itself. What a concept to understand that the black hole which maintains a galaxy is merely sowing the seeds of solar systems to fuel its appetite. In consideration of this analogy, our humble existence might be viewed as nothing more than our participation within the consternation of this galactic bacterium.

The nature for the containment of our universe, or what has traditionally been termed as the fourth dimensional state of our continuum in Time and Space (or alternatively the eighth dimension), provides for a balance of forces both large and small. The large (or strong) force continually try to exert its individualized force via its positive matter (or positive mass density) in gravitational acceleration, while the small (or weak) force synergistically fills in all the empty spaces via its negative matter (or negative mass density) in gravitational deceleration. But this dimensional state of affairs is little more than the transient condition for our bubble of containment in Space-Time. Such dimensional transitions are fluid, transmogrifying the state of existence from one dimension to another. From our imagined state of singularity at the time of creation (or the Big Bang) through to our present state of the universal condition and into whatever the next dimensional transition holds for us, our energies form impressive effigies within the preternatural ether of the altering Space-Time continuum. Is there life after death? I guess that all depends on what you define as life. Perhaps death is merely the twelfth dimensional transition of life within six dimensions of Space and six dimensions of Time.

It is this magic of six dimensional Time that has captivated science fiction enthusiast with a flare for imaginative abstraction. For the speed of light is only a pale imitation of what has been suggested as Time travel. Such travel in only five dimensions of Time merely relates the ability to distend Time in Space to express the illusion of Time Travel. For real temporal displacement, one would have to travel in six dimensions of Time to purport the ability to leapfrog in Space instantaneously. But even travel within a continuum involving five dimensions in Time and only four dimensions of Space expresses a real parody of an ability to measure Time outside of an expectant eighth dimensional Space-Time reference. Would the evolution of matter just skip the ninth dimension in favor of a more balanced tenth dimensional condition of existence: Life within five dimensions of Space and five dimensions of Time? In movies like '2001: A Space Odyssey' or 'Slaughter House 5' and science fiction stories like 'The Man Who Folded Himself', there is a clever presentation of what I imagine of this continuum in six dimensions of Time and four dimensions of Space. Only problem is that the physical interpretation of mass in these stories is more relative to a traditional fourth dimensional condition. Still it wouldn't be much of a story if all matter was compacted into one mass density. So perhaps like death, this distorted tenth dimensional condition is a representation of our consciousness in all instances of Time; Life without end with nothing but the solace of our thoughts.

Chapter 8

"The Infusion of Relativity"

Prologue

In the previous chapter, 'How', a subsequence of the interrogative 'Why,' was examined as the implied convention of purpose towards expression. As it has been presented, its expressed purpose was represented as an expression towards 'being' within the interfusion and interdependence of the bounded infinities of Space and Time for existence. It is this dimensional abrogation of a previously infinite void that transitions the condition of existence from within an abyssal state of non-existence. As the 'how' of existence has revealed an existence from non-existence, we continue with an examination of the forces by which matter persists as the fractionalization of this process to further bind infinity from within evolving dimensional transitions.

Let us reexamine what we know about gravitational interaction: theoretically, as mass is continually accelerated across Space in Time, it increases in density by virtue of its increased acceleration. Its increased density is relative to its displacement, or invasion, of dark matter which is permitted via its mass expansion. The decrease in the commingling of dark matter provides for a measured decrease in the surface area upon which a mass' density is defined. As this increased acceleration of mass in motion is then further altered to accelerate within its own Space and Time, it will eventually attain enough density to achieve an inward gravitational acceleration greater than its motion of outward acceleration. When this happens, the outward acceleration of mass in motion is internalized to spiral back in on itself, by virtue of its greater inward acceleration, causing it to gravitationally collapse in upon itself. Herein this force of gravitational collapse, interfusing both the inward and outward acceleration of its mass, engenders a greater volume of its influence upon the surrounding displacement of Space in Time. Via its increased acceleration, its altered condition is forced to balance itself in an inversion of its Space in Time. The displaced dark matter collapses in on the mass itself. In other words, the mass itself is literally turned inside-out, due to the interfusion of dark matter, and its volume of influence grows exponentially beyond the limits of its previous container.

Such an inverted mass, having warped Space and Time by turning itself inside-out, is no longer considered to be exhibiting an outward motion representative of mass in motion; rather it imagines an inverted notion of motion liken to mass at rest and is similarly defined by its increased influence within the Space-Time continuum. Upon the increase permittivity of dark matter without the mass that provided for a decreased measure in the surface area upon which a mass' density is defined, the mass is isolated and compressed. Herein the commingling conditions of gravitational acceleration and deceleration, a compression of mass within an increased concentration of dark matter is overwhelmed by their difference. The energy of the mass is released against a rapid decompression in its degenerative pressure of repulsion. The rapid decompression is expired upon its "absence of density" in a rippled effect that immediately accentuates the condition of this difference between its energy and the surrounding medium of dark energy.

Its accelerated condition once again becomes exponentially greater than it was before, dimensionally balling up Time: i.e., liken to twisting a rubber until it can do nothing else but form a ball. Some theorize of this condition that Time has greatly slowed down or stopped relative to the Space in which it exists. Subsequently where Time has been balled up or stopped, Space is not being measured so much by its gravitational acceleration but rather by its gravitational force in terms of its measured matter under pressure, or foci of compression. This being true, then the gravitational condition that exists, relative to the volume of Space in which it has influence, is representative not only by the compression of available mass for such a gravitational center but also by the conditions through which is was engendered. Ergo the condition of a black hole is more representative of 'maximum Space' and 'minimum Time'; wherein a black hole is simply the measured force of massless density upon the interfusion of mass and dark matter. Indeed, the hole is real as there is nothing at its center but a Space-Time condition that only allows mass to energy conversion.

Using this logic then, even the less dense matter that extends far beyond the nucleus of the gravitational center must also play a factor in the full measure of a considered mass. Further still, this extended matter would also play a factor in the full measured influence of a mass' density for such a gravitational center. The resulting condition of the forced gravitational acceleration however would lack cohesion and begin to tax itself upon the extended system of its mutually beneficial mass to energy conversion. The system must feed on the mass of its extended system in an effort to retain the neutral balance of its gravitational center. Yet in doing so, it depletes the foci of its gravitational condition. In its imbalance, the extended matter then distorts its relative degeneracy into an alignment of its expansion along a rotational spin that has become increasingly active due to its entropic condition.

While this condition increases the gravitational acceleration of the mass towards it expansion, it further accentuates the rapid decompression of its energy along its poles in a wake of increasing gravitational deceleration. As the extended mass is increasingly accelerated into this entity of unbalanced gravitational influence in distortion of its own degeneracy pressure, it is forced into a state of imbalanced that further upsets the foci of its neutral gravitational center and only adds to its problem. Like a cancer, it can only support its growth as long is there is viable mass to feed upon. In example, the black hole at the center of the galaxy could not exist without the extended matter and energy that makes up the body of the galaxy. The overall volume of attractive influence is actually defined by the accumulative mass required to amalgamate a layering of these mass densities towards engendering its focus of a gravitational center. This is contravening to an explanation of galactic formation as an expression of the independent embodiment of a black hole that only attracts and captures mass relative to its gravitational interaction.

Consider the layered and rippled conditions of the interaction between opposing gravitational forces. Where a gravitational deceleration continues to dominate, a potential insistence is inspired such that the pre-imagined ideation of existence, in structure of its atomism, is all but diluted towards a being in/as/of nonexistence. In this way, the mass' density is transferred to an ever expanding compass of kinetic expulsion in an ever accelerating universe. Although it is difficult to abstract such a pre-existence within/from nonexistence, it can be measured upon its condition as the boundary of 'maximum Time' and 'minimum Space'; where 'all Time' is measured in tenure. Herein at the edge of identity, where there is no mass for lack of spatial dimensioning, is an entification of Time as the reveled revelation of being. It is this beginning of Time that represents the last bastion of identism in which some religious philosophies refer to as the notion of a Heaven.

Again consider the layered and rippled condition of the interaction between opposing gravitational forces. This time however, consider their contention where gravitational acceleration would continue to dominate. Where a gravitational acceleration continues to dominate, a kinetic subsistence is inspirited such that the pre-imagined ideation of existence in structure of atomism is concentrated towards a being in/as/of existence. In this way, the mass' density is focused in an ever compressing volume of potential impulsion in an ever accelerating universe. An abstraction of this ideated existence can be measured upon its condition as the boundary of 'minimum Time' and 'maximum Space'; where 'all Space' is in measure of tenure. Herein at the focus of identity, where there is no absence of mass for lack of temporal dimensioning, is a realization of Space as the dispirited revelation of being. It represents the embroiled involution of identism in which some religious philosophies refer to as the notion of a Hell. In the ingratiation of such religious idealisms, one might ask if they themselves are contributing toward the universal precepts of perpetuation in greater deceleration or greater acceleration.

This speculates much as to the age of the Universe because its presents the entification of Time and the engendering of matter as a random collection of chaotic conditions influenced within the Space-Time continuum. It purports a sense of ideation in which there was no one Big Bang, but rather many such transversely dimensional events. Since each of these Big Bang events could be inspired independently of each other, it promotes the idea that pockets of such gravitational conditions could spring forth within the chaotic amalgamation of matter focusing toward density by volume. As a result of this condition of introspective interference in Space-Time, loose matter and/or mass outside the more physically defined nucleus of gravitational center would be purported to influence a Big Bang without actually being the focus of its impetus. With this understanding it would not be necessary for all of the mass in the universe to amalgamate into one universal Big Crunch, causing one Big Bang, to explain a viable universe. Rather it is very unlikely that just one such event would define the viability of a single universe. Instead it is likely more probable that multiple Big Bangs would increase the viability of the universe by randomly spreading matter out enough to maintain a more uneven distribution with which to discourage a universal center. In fact such randomness would probably improve the stability of the universe; somewhat like a boiling pot trading off hot points, it would keep the universe from obliterating itself completely. Therefore as long as the precondition of the initiated dimensional incursion persisted, the universe would be defined via its Space-Time continuum.

Although such localized events, like a Big Bang, happened in an instant from our dimensional perspective, the instant is relative to the warped nature of its dimensional inversion of Time over Space. Maybe what is envisioned as this one Big Bang is just one event in a somewhat isolated dimensional pocket of a more robust universe within infinite possible degrees of Space and Time. This may still leave the mystery of what condition might have triggered the implosion of an imagined Big Bang phenomenon. Although localized, it might be something like a total extirpation of Space and Time discreating with the force imagined of a matter-antimatter encountered upon a dimensional incursion of Space and Time: i.e., the dimensional transmutation from one combinational Space-Time continuum to another. Realizing such an encounter can be envisioned as the forcing of a brief perlustration into the presageful abysm imagined outside the bounds of Space and Time, somewhat like the presentment of a dimensional windowing into a being in nonexistence. So where does one go from here? After all, as considered in volume one of this book collection, everything (existence) starts with, or starts from within, nothing (nonexistence).

A Universal Centric Singularity

In consideration of a measure toward ever increasing acceleration, remember everything is by degrees. Even unto the resultant product of such an enfolding event, the outcome is dictated by the rate of the rapid decompression of its decelerated repulsive force. It not only speculates as to the amount and rate of energy released against the attractive force of increasing mass being accelerated, but it also portends the capability of momentarily trapping the escaping energy against its own rate of increasing deceleration, resulting in an implosion: An explosive force caused by violent compression of greater mass acceleration. This rate of rapid decompression is heavily dependent on its ability to attain greater acceleration. Since measured mass density dictates a measure of greater acceleration, then greater mass density provides the ability to increase the rate at which greater acceleration is attained. This is why there is a varied range of resultant products and/or events evoked from this simple mechanism; from White Dwarfs to Quasars to Black Holes to Supernovas to Big Bangs. Yet if the Big Bang is not the beginning/center of our universe, then what is?

It has been postulated that there is no real center to the universe; and this would be a true statement if one attempted to define it via any physically known measures. However a universal center can be abstracted in the ability to qualify a concept in physics that similarly theorizes the notion of a 'singularity'. A singularity is the considered womb in which existence was engendered and the considered primer for the conceived Big Bang. A singularity is sometimes imagined as the considered center of the universe, or the initial state of the universe from which all matter was defined. Rather a singularity is simply the state of a Space-Time's conditional inability to be measured: Infinity. While a singularity contests the congruent laws of physics, it should not be viewed as a break down in these scientific precepts. Rather it is the inability to interpret such infinite measures of mass, energy, or force from inside the confines of our current dimensional reference.

Instead, I would propose that the universal focus is more of a persistence in the continuous diffraction of an incipient condition toward a convergence in Space-Time; the suspended continuum of 'being' in a medium of singularity for an existence from nonexistence. More easily understood likened to its measured potential for existence, this 'being' pronounces a state of non-existence that animates a diametric condition that belies the very nature of its abyss. And yet it is this very absolution from the abyss that provides the connective insistence towards a beginning; a beginning from which to initiate an origin and generation towards a more animated existence. Where NE is non-existence, EO is existence with opposition, and EA is existence with affinity:

NE ------------> EO
NE <------------ EA

NE -------------------------> EO
NE <----------- EA <----------- EO

(attracts)
EA ---------> NE ....Existence takes being in nonexistence;
(attracts)
EO ---------> EA ....Existence takes being in identity;

(repels)
EO <--------- NE ....Existence takes identity in being.

That is, if the notion of a singularity is defined as outside of the dimensional aspection of our universal condition, it is more a specter of No Time and No Space that mirrors the refraction of 'minimum Time' and 'minimum Space' against the complementary diffraction of 'maximum Time' and 'maximum Space'. Upon this proposal, the universal centric singularity is the omniscient omnipresence that divines inspiration in 'being' from the sentience of its own infinity.

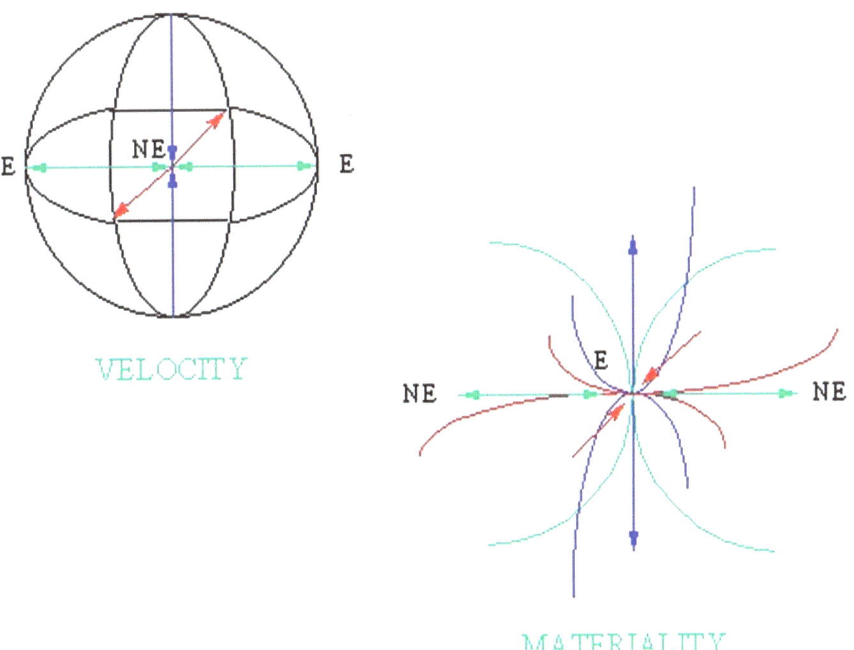

Figure 8.1

Evermore, it is this binding concentricity that imbues the universe within a relativity of Space-Time; the precondition of the initiated dimensional incursion in which the reciprocal forces of gravitational acceleration and deceleration occur for 'being'. So there is and isn't a center of the universe depending on your model of abstraction for its relativity. It is not a center which can be viewed, or even measured, as its very reality is subject of debate. It is this persistence that purports, in our dimensional measure of being, a flow of influence to and from center. Thus the notion of a universal singularity is derived, not from its focus of centricity, but from within its ambit of containment. It is not a fixed location in Space, but rather a fixed condition in Time. These are the conditions of our creation and termination.

Yet, there should be no one Big Bang to narrowly divine the 'being' of the universe. Indeed, this definition should manifest and embrace an understanding of multiple universes existing within a more supereminency of all universal conditions. However, rather than deal with such an abstraction, one merely needs to agree on the boundaries that outline a unique dimensional incursion. Remember everything exists by degrees. As such, matter appears to have a narrow range in which states of mass and energy can exists within the Space-Time continuum. And the random nature of such localized events as Big Bangs (or dimensional incursions) and Big Crunches (or dimensional excursions) can only provide for pockets of discrete events within the infinity of the whole universal being. Thus, such events have no real significance to any measure in the overall age or size of the more supereminent view of a universal 'being'. It is also questionable how much weight can be placed on calculations for universal models that continue to utilize a closed system in measure of mass to predict whether the more supereminent universe is expanding via a Big Bang or ultimately contracting towards a Big Crunch. Rather the random nature of our localized pocket exists as a result of its condition within an ever dimensionally transmuting Space-Time continuum. For remember, everything only exists via the degree in which Space and Time converge.

Bounding Infinity

Returning to a recollection of the interfusion and interdependence of these conditions for creation, what considerations have these protractive obstrictions made in respect to their dispositions? The pervious perusal of previous reference relates a magnification of inflective measure to a maximum constant velocity in an introspective purview of focus toward a reflective measure in perception of a maximum variable velocity: A focus of maximum deceleration from maximum acceleration (see Figure 7.3). Even as a point gives rise to a line/curve and/or sphere in the irrepressible irruption of its precision to irradiation, a concave variability of velocity focuses on a resistance of maximum deceleration to maximum acceleration (see Figure 8.1). In view of a measure to maximum distance, the end point of perception, one may now pursue the ambit of existence towards an infusion of relativity in a newly acquired, fourth dimensional invigoration; a fourth dimensional invigoration extracted from the accentuation of its percipience in a resonance of its resoluteness to such interfusion of interdependence in these maximums. These maximums of measure combine to define a renewed appreciation of energy in both state and form essentially inherent to all being: The original four fundamental forces of creation. Such is it that difference subsists as the suffusion of persistence in sufferance. As earlier introduced, existence is the negative of non-existence, where existence knows being in attraction and repulsion. One further remands upon their motions in complement, as well as in contention, seeking to design the very fabric of their insistence in form of their difference. The objectivity of their predilection to combination for relativity, these four fundamental forces are:

1. Maximum Acceleration of Maximum Distance
2. Maximum Deceleration of Maximum Distance
3. Maximum Acceleration of Maximum Resistance
4. Maximum Deceleration of Maximum Resistance

Herein lies the qualification of measure to form in the incipient characterization of an inextricable infirmity that belies the infinity of its nature. Recalling figure 7.3 as the earlier delineation expressing an insinuation of this combination, one begins to animate a view of energy to its relative entification in the definitions of state to its fourth dimensional invigoration:

1. Maximum Acceleration	Minimum Time	
Maximum Distance	Maximum Space	NE
2. Maximum Deceleration	Maximum Time	
Maximum Distance	Maximum Space	E
3. Maximum Acceleration	Minimum Time	
Maximum Resistance	Minimum Space	NE == E
4. Maximum Deceleration	Maximum Time	
Maximum Resistance	Minimum Space	E =/= NE

Thus, invigoration via condition in combination defines an interim of interaction, among the relegated states of inception, bringing about and necessary for our traditional view of a fourth dimensional circumscription.

This, the comparable compendiums of two concernments (e.g., non-existence and existence) in sympathetic apotome (e.g., NE == E) to an apophasis of appropriation (e.g., E =/= NE), proved to be my greatest vexation. Abstracting its apodixis (or absolute certainty beyond the contradiction of its symbiotic association), probated as the ambiguous nature of its inception, I worked at it countless times in a variety of ways. And yet, sometimes to find the right answers, one has to find the right questions. So it was with me as I asked myself: What are these newly combined postures suggesting, via definition in adjunct, to be their relative actualities in existence? I answered such in this way: The perception of each is dependent upon the reality of their activities in energy or mass, via definition, to the apposed appositeness of their adjunctive synergism. Such is it that:

1) The relativity of Maximum Acceleration of Maximum Distance is the expansion of mass (Space) for energy without resistance;
2) The relativity of Maximum Deceleration of Maximum Distance is the contraction of energy (Time) for mass without resistance;
3) The relativity of Maximum Acceleration of Maximum Resistance is the expansion of energy (Time) in resistance to a constant of Space;
4) The relativity of Maximum Deceleration of Maximum Resistance is the contraction of mass (Space) in resistance to a constant of Time.

Remanding a preceding characterization of these terms unto the specialized complicity of their compliance to such binary combinations, I further sought to define their interdependent individualities upon expected principles that could repute their occlusion in polarity. Purported as the primary provocations to provenience:

A. Minimum Time (Maximum Acceleration) defines a view in the negative entropy of maximum order;

B. Maximum Time (Maximum Deceleration) defines a view in the positive entropy of minimum order;

C. Maximum Space (Maximum Distance) defines a view in the positive enthalpy of maximum potential energy and endothermic properties;

D. Minimum Space (Maximum Resistance) defines a view in the negative enthalpy of minimum potential energy and exothermic properties.

Pursuant to a presentment of this proviso, retouching upon a perpension of their adjunctive postures in apposed appositeness further supported this association as to their relative actualities in existence; to which:

a) Energy without resistance, as in the relativity to Maximum Acceleration of Maximum Distance, suggests a foundation for potential energy (origin);

b) Mass without resistance, as in the relativity to Maximum Deceleration of Maximum Distance, suggests a foundation for attraction as in a proton;

c) Energy in resistance, as in the relativity to Maximum Acceleration of Maximum Resistance, suggests a foundation for kinetic energy (neutron);

d) Mass in resistance, as in the relativity to Maximum Deceleration of Maximum Resistance, suggests a foundation for repulsion as in an electron.

Note: 'origin' outlines an unperceivable state suffusing existence for invigoration.

A comprehension of this proposal would seem to foster the pre-existence of a mass-like quality within the established euphemism of such fundamental forces. The logic of its facilitation to a degree of inception would then necessitate the infusion and/or implied concrescive conglomeration of such mass to an instance of materiality based in the predilection of its randomness unto participation; yet this is not the case. This calculated remonstration of non-existence merely purveys a sophism (or cunning contrivance) of relativity that could only be founded upon the fanciful pragmatism of Space-Time warps that are purported to maintain an existence of perpetuity within the ethereal matrix of the universe. However, one shouldn't have to distort the precision of relativity to preserve the conceptualism of such esoteric theories.

Rather, relativity should follow the insistence of an inception in the non-existence based on the applied exemplifications of known empiricism. Indeed, this relativity is the relegation of an abysmal infinitude that prompts the persistence of inception unto existence. This, the negativity of an existence within that ubiquitous quality of positivity held in an atavism of non-existence, in no way supports the paradoxical parody of comparison between matter and antimatter as the inherent prodigy of creation that infers a prefabrication, and subsequent prepossession, of a symmetrical sense of precision within the universe rather than the asymmetrical subsistence of their individual instance: i.e., Mass is either matter or antimatter, not both, as prior expostulation to this notion of parallelism in deviance of spacial relativity has pointed out or supported. Relativity is not predetermined to follow a homogeneous and isotropic pattern in the evolution of the universe. Rather it is the random nature of its apostasy that provides for the eventual entification of our dimensional continuum.

Recapitulating for combination in design of a theoretical presentation of the constitutional developments of an atom, the possibility, and therefore the probability, of a creation and evolution, or of a creation in evolution, can be more fully appreciated. Note of the original, four basic interactions:

1) (Origin); + (positive) and − (negative) and or + −

Maximum Distance	Positive Enthalpy	=	Maximum Energy
Maximum Acceleration	Negative Entropy	=	Maximum Order

2) Proton; + +

Maximum Distance	Positive Enthalpy	=	Maximum Energy
Maximum Deceleration	Positive Entropy	=	Minimum Order

3) Neutron; − −

Maximum Resistance	Negative Enthalpy	=	Minimum Energy
Maximum Acceleration	Negative Entropy	=	Maximum Order

4) Electron/Positron; − +

Maximum Resistance	Negative Enthalpy	=	Minimum Energy
Maximum Deceleration	Positive Entropy	=	Minimum Order

At first reflection, groups 1 and 4 would appear to cancel themselves out. This was a problem for me until I examined the spontaneous decompositions of atomic nuclei. It suggested to me that the neutrons and protons of the atomic design must be somehow interchangeable given the proper circumstances and conditions. Subsequently, I patterned this hypothesis in the following representative relationships:

N (Neutron) = p (proton) + e^- (electron)

or = p - e^+ (positron)

P (Proton) = n (neutron) + e^+ (positron)

or = n - e^- (electron)

Substituting the positron for its reciprocal counterpoise, represented upon the notion of its subject stupor to an electron loss or 'hole', I incorporated the study of subatomic quarks for completion of a viable, working model from which both, proton and neutron, endure as the synergistic packing of electrons (e.g., negative or positive), as shown below:

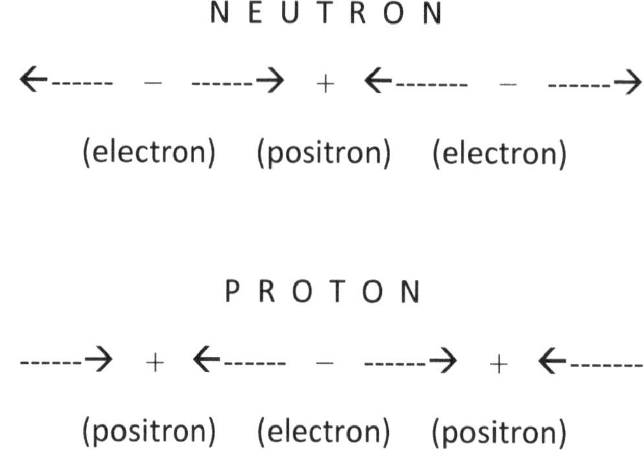

Arrows are provided for the expletive illustrations of expected polarity within the exosmotic activity generated via the magnetic monopoles expiscated, and are suggestive of the synergistic packing of electrons for greater energy whereby groups 1 and 4 embody the insitu sobriety of their recognized subsistence in forming the predominant tenacities that ingratiate protons and neutrons.

Thus, this theory intrigues an incipient comprehension with which to enthrall; what implications would the design of protons and neutrons have on the mathematics, mass, or synergistic value of electron packing and subsequent polarity? Further, how might the positron, or lack of electron, be considered in the make-up of such mass? One possible promise follows:

Where a neutron ~= 1.0087 (at. wt. scale) and

a proton ~= 1.0073 (at. wt. scale) and

an electron ~= 0.00055 (at. wt. scale)

Then $2(e^-) * 1(e^+) \sim= 1.0087$ and

$1(e^-) * 2(e^+) \sim= 1.0073$

Noting '*' as unknown functionality in the factoring of its value.

In review, the recognized interdependence of these four basic interactions in the design of an atom reveals both polarity and copulation of an interfused binomial existence toward the further evolution of energy in mass:

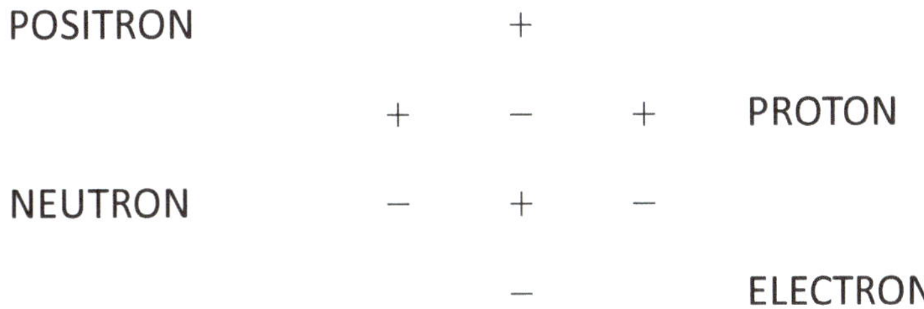

One now assimilates an articulation of mass to atomic weight, via its increasingly inextricable complexity in a furthered qualification of its fourth dimensional interrelations, as the circumscription of energy to a field of order imbibing the antithesis of either amity or enmity to the animate anonymity of entity; e.g., the associative symbiotic condition of its inherent nature. Thus, the proton relates the difference between the Maximum Deceleration and the Maximum Acceleration, newly incurred, of Maximum Distance; The neutron relates the difference between difference between the Maximum Acceleration and the Maximum Deceleration, newly incurred, of Maximum Resistance; Whereby the electron becomes the visible indulgence of contracted mass that, being expendable to the corporate structure of the atom, can be expended; While the positron relates, rather, the dissolution of such energy or expanded mass.

This enigmatic choreography of determination – passive and active, attraction and repulsion, female and male, if you will – are forever interlocked into an interrelationship invoking an involution in evolution. It is from this point of maximums that evolution of Time and Space, as well as of future and past, is predetermined upon its eventuality, long before its perception. Copulation, here, is then redefined as the interfusion of difference via polarization. One can also imagine the susceptibility of such ephemeral expressions to the incipient conjuncture of its ideorepulsive natures upon the incorporation of these magnetic monopoles. An incubation for identity begets this necessity of affinity to affection for which to avert expostulation (or expectant annihilation) of being upon impercipience. And yet, such exigency exhilarates an expedient existentialism with which to exert an exhaustion of perceptibility in its own egotism. However, this deliberation is but a philosophical overview of previous presage to the precocity of purpose in being.

Returning to an examination of atomic development, one now unfolds the miracle of evolution upon the four combined fundamental forces to substantiate the substructure of energy in aggregation of resistance. This conception remands the prior analysis of existence via delineation for improved comprehension in which shall be utilized the distinguishable predesignations of their postulation (see figure 8.2).

Figures 8.2 - a,b,c,d

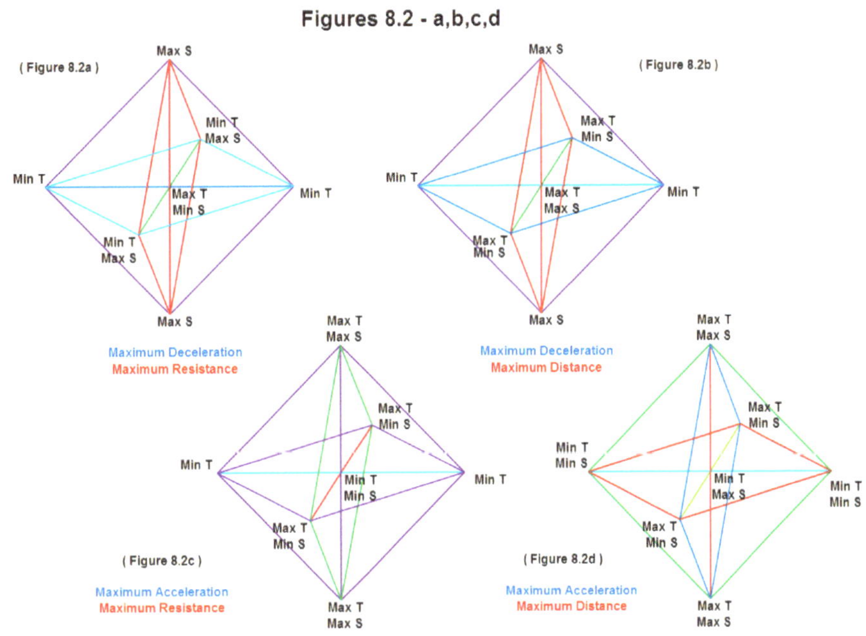

Here one can deduce, from a consideration of the ideas demonstrated, that deviating from a constant velocity, the abeyance of least resistance, abets polarization of energy in which such polarization induces that quality of entity to production.

And yet, this interdigitation presents a reminiscence of a prior analysis in perspective of polarity which seeks to allay the abstruseness of interfusion and interdependence to these maximums in design of their fixed relativity within the Space-Time boundaries of existence. Note, all four illustrations in figure 8.2 are comparable to the first four steps depicted of figure 4.14 (as previously presented from 'The Evolutioning of Creation, Volume 1') in the 'immersive liquid inversion' progression, to which now would seem to have been inappreciably articulated.

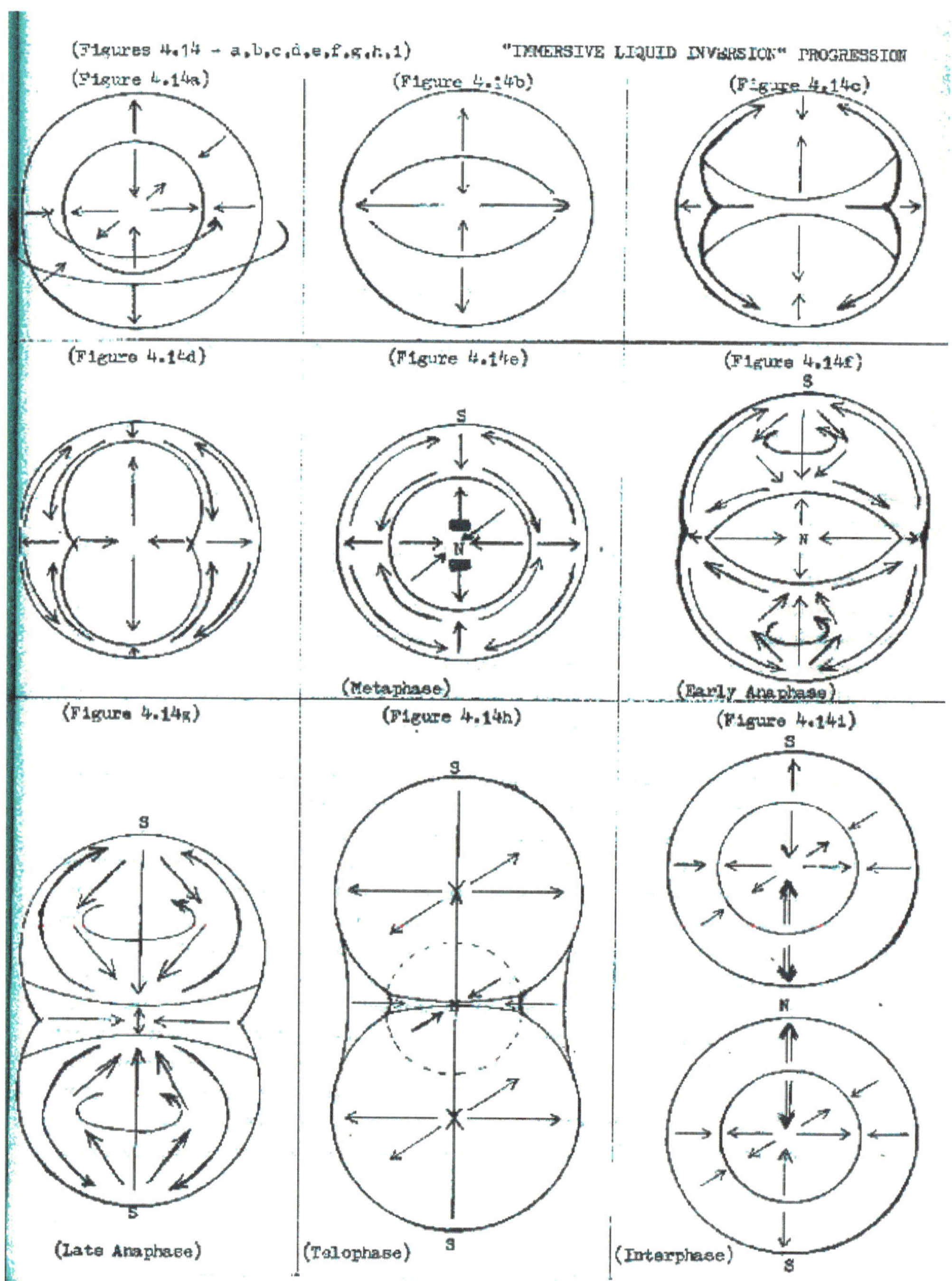

Of the last illustration in figure 8.2, coincidentally representative of the biologically recognized condition of metaphase in cell division, one can imagine upon the distinctiveness of Time and Space (see figure 8.3).

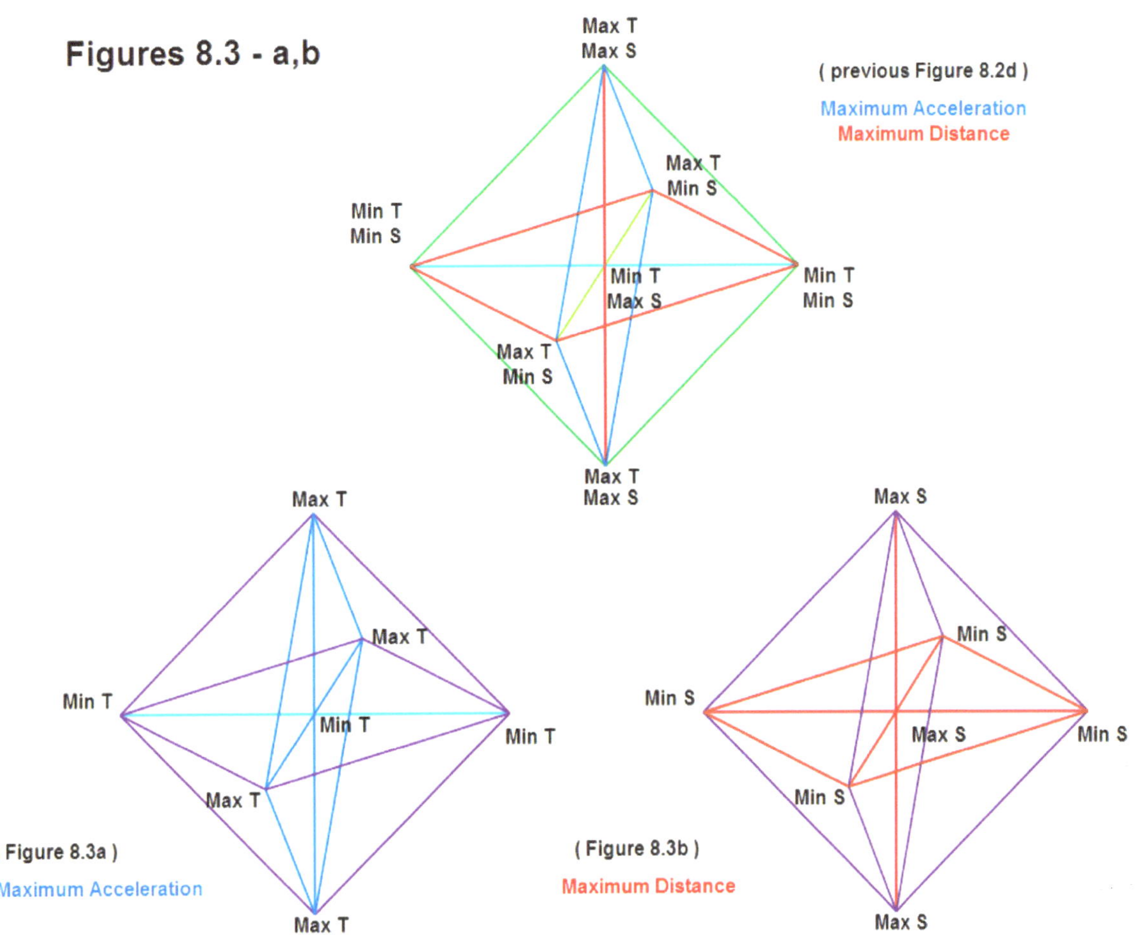

Adventuring a step farther in this examination of energy development, I wish to impose a facilitation of atomic components with which to imbue the delineations of Space-Time within a concert of substructure in energy. Now, look again with me as I begin to picture the evolutioning of energy as one slow, lethargic, step by step process into an analysis of cell division (as envisioned via the assimilated immersive liquid inversion) starting with figure 8.4. Stop here for a moment, if you will, to prescind (or abstract) the synchronal synonymity (or identity of nature) of this special hypotyposis (or work picturing) that seeks to pioneer a predisposed pretermission in the logical positivism of evolution.

If you haven't hypothesized unto a conjecture of this premise to intent, then pause for this moment to recall the expected nuclear compositions of elements and their predisposed atomic and nuclear substructures. What I am theorizing toward is an evolution of energy for mass in the nature of atomic and molecular orbital diagramming by means of their innate prepossession to a fixed relativity within the Space-Time extent of existence. By combining the components of these assimilations presented and demonstrated, one can continue to portray the expected neutron and proton relationships within a fourth dimensional perpension in the building of substance towards a logical character of existence within the universe: A philosophication unto the mode of physiosophy (or a doctrine concerning the secrets of universal nature) and/or pansophy (or a system embracing a universal wisdom).

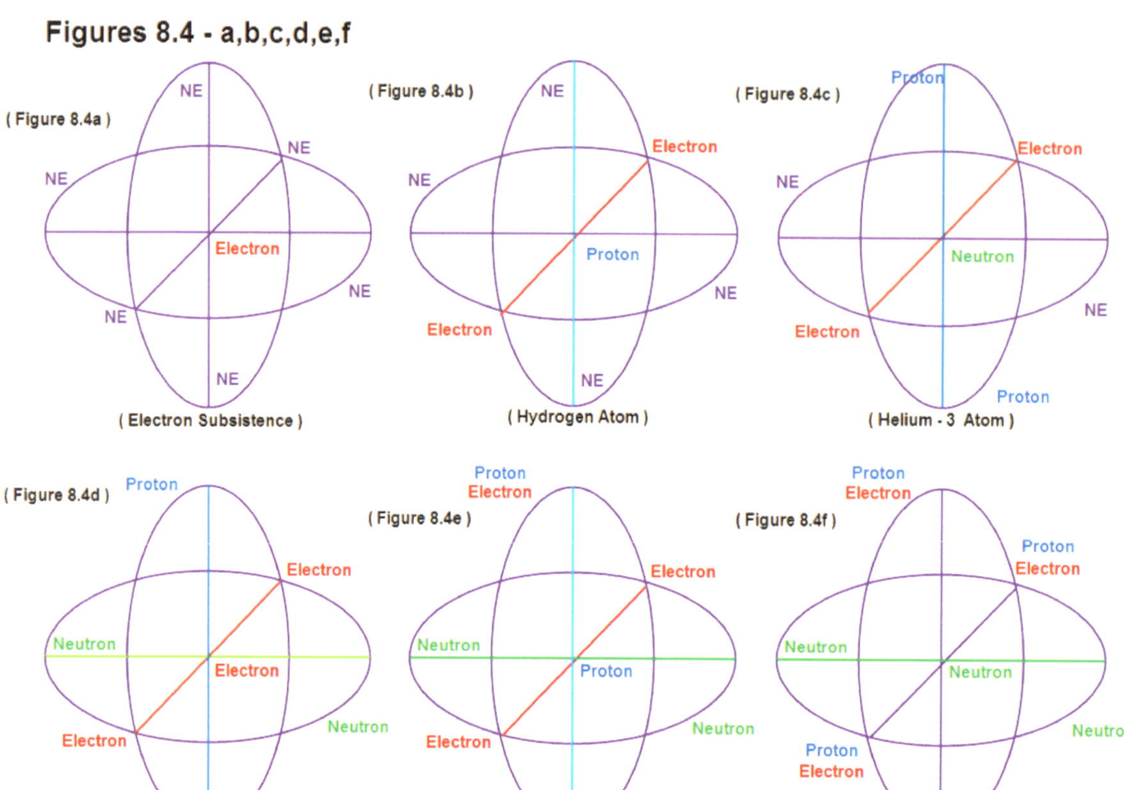

This illustration in the study of atomic and molecular evolution of energy, indicative of the hydrogen molecule in figure 8.4c, precedes a pattern idealizing its idiomatic ideology. Thus, the fourth illustration (see figure 8.4d) brings about a renewal of this pattern in the emulous imitation of the configurations presented in figures 8.1a through 8.1c. This then relates the components to the atomic structure of helium on up to the sixth illustration, displaying an aspect of the beryllium atom (see figure 8.4f).

E	1E	(electron)
E P E	2E,1P	HYDROGEN-1
E P N P E	2E,1N,2P	HELIUM-3
E P N E N P E	3E,2N,2P	2 HYDROGEN-2
E P N E N P E	3E,2N,2P	HELIUM-4
E P N E P E N P E	4E,2N,3P	LITHIUM-5
E P N E P N P E N P E	4E,3N,4P	BERYLLIUM-7
E P N E P N E N P E N P E	5E,4N,4P	BERYLLIUM-8
E P N E P N E N P E N P E	5E,4N,4P	2 HELIUM-4

E P N E P N E P E N P E N P E	6E,4N,5P	BORON-9
E P N E N P E P N P E P N E N P E	6E,5N,6P	CARBON-11
E P N E N P E P N E N P E P N E N P E	7E,6N,6P	CARBON-12
E P N E N P E P N E N P E P N E N P E	7E,6N,6P	2 LITHIUM-6
E P N E N P E P N E P E N P E P N E N P E	8E,6N,7P	NITROGEN-13
E P N E N P E P N E P N P E N P E P N E N P E	8E,7N,8P	OXYGEN-15
EPN E NPE P N E P N E N P E N P EPN E NPE	9E,8N,8P	OXYGEN-16

Stopping again, here, to preponderate upon another pretermission in the design of atomic and molecular substructures, one should ask, where have all the electrons gone. The answer is that they have not disassociated themselves, but that the model originally utilized in ought of their propinquity is not designed to picture them. And therefore, the configuration must be expanded to accommodate its expedient growth. However, note that the foci of the presented models of atomic and molecular development are inverted converse to its order in expected, fourth dimensional, orbital energy diagramming. This, the conceptual ideology of the synergistic packing of energy towards its evolution and continued development into a perceivable palatability as matter, is achieved via a fifth dimensional relationship in relativity pursuant to the varying and modifying of materiality accordingly to velocity, pressure, and concentration in concordance of shape. Shape becomes an increasingly growing concern to a being in resistance because of its transitive and intransitive transitions in its development toward a more complex, evolutionary existence. Further, all matter is but energy, and it is how each being perceives such that defines their existence. That is to say, just as all matter is a degree in perception of energy, our being and its environment, our universe if you will, is dependent upon our perception of this energy, predicating a sixth dimension in existence qualifying function.

As one might have concluded of the preceding determinations of existence to energy, a dimension, via definition, prefabricates a convention of being into an expression of qualification in existence; whereby each succeeding dimension further modifies the purview of existence within he qualification of being. And so, of the last three figures presented (see figures 8.2, 8.3, and 8.4), incorporate an analysis of the transitional stages of energy heretofore approached in ought of the four basic interacting forces of energy combined to lend animation to being in existence.

Such is it that for its inception, I shall remand and revise a review of that binary synergism of Time and Space giving direction to velocity in a manner reminiscent of the agreed notions of polarity from conception in creation (see figure 8.5).

Figures 8.5 - a,b

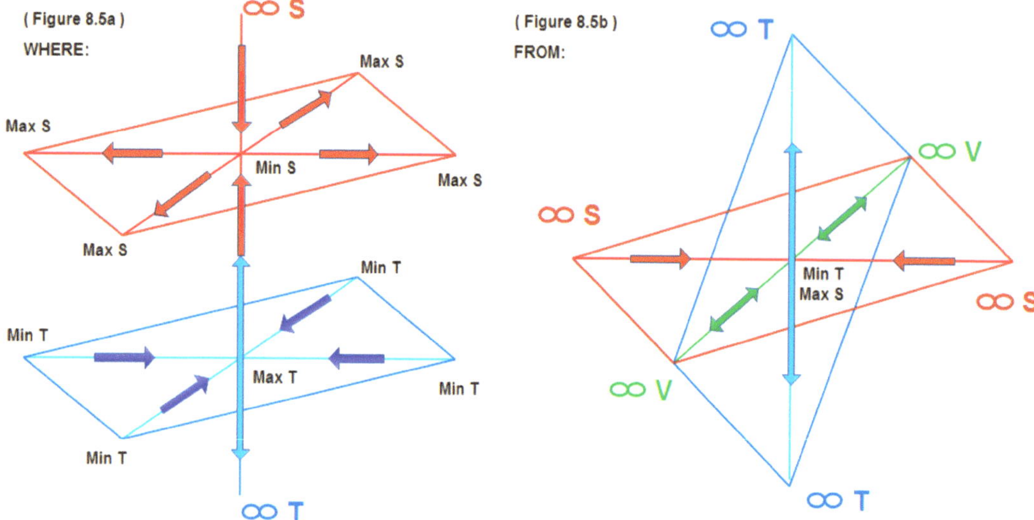

The, the planes of Space and Time overtly drafted, having brought about a consummation of existence, are drawn into perpendicular attitudes to each other so as to nurture or focus this velocity into a development of motion via its implied resistance. This, the attitudinizing of velocity in focus, constitutes the 'metaphase' condition of prior examination, whereby the primordial predestination of its precipitous preference prefixes a prepossession of its probation within the promptitude of its incubation. As provisioned of figure 8.5, the imminent cultivation of Time and Space toward the auspicious autonomy of function in velocity propagates the immanent precocity of pregnancy within the culmination of evolution in relativity upon the ideorepulsive nature of its inception. Motion is resistance is, then, expanded in Time and Space (or Space is expanded in Time) upon the inversion of velocity into a being in kinetic mass enrapturing a condition of innate inertia, a being in potential energy, such as follows in figure 8.6 (see figure 8.6).

Figures 8.6 - a,b,c

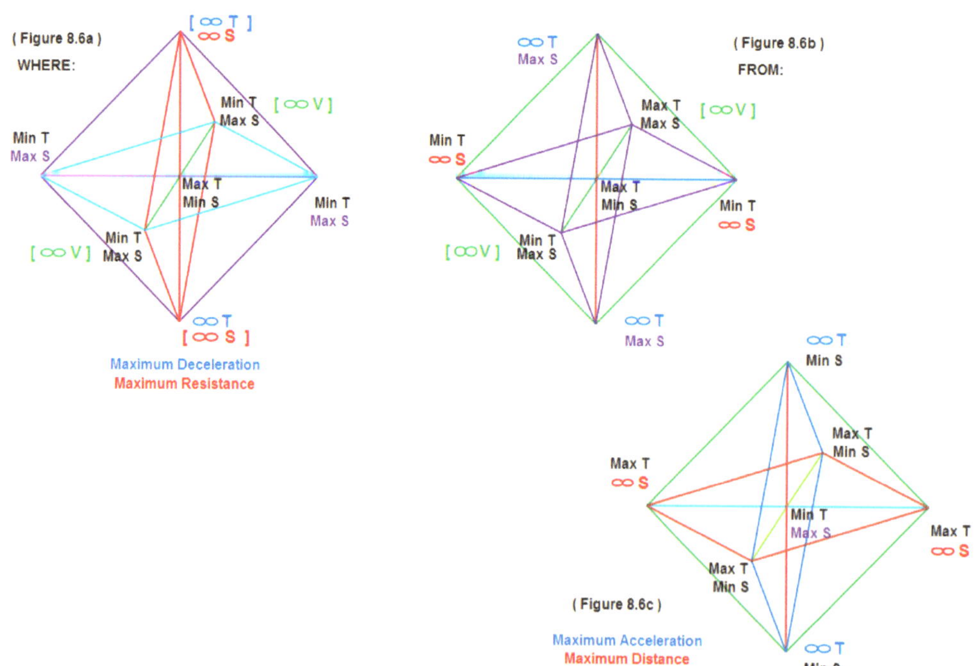

This inversion of being was first proposed as the divinization inspiring existence from within non-existence, or that instantaneity instating event from within the implexion of relativity. Here, one can more aptly imagine the professed birthing of relativity in perception of materiality commensurable to a confirmation of mass, as discussed in chapter 2 of 'The Evolutioning of Creation: Volume 1', wherein it is the tangent that becomes the foci of measure for the inception of perception in matter. Here, the 'Evolutioning of Creation' imbibes the imbrications of difference in velocity toward the incipient institutionalizing of materiality. This evolutionary process of energy development has gone from maximum potential energy to minimum potential energy through the promotion of energy into a state in kinetic mass, just as a complementary process will go from maximum kinetic energy to minimum kinetic energy through the promotion of energy into a state in potential mass; Energy begets energy as the incessant motor of life prevails.

As this whirligig of atomic and molecular evolution continues upon these four binary conjunctive primes in facilitation of energy to development, note the formation of potential mass, mass without resistance (proton), that bears the quality for enceinte (or with unborn offspring) induction upon its inception (see figure 8.7).

Figures 8.7 - a,b,c

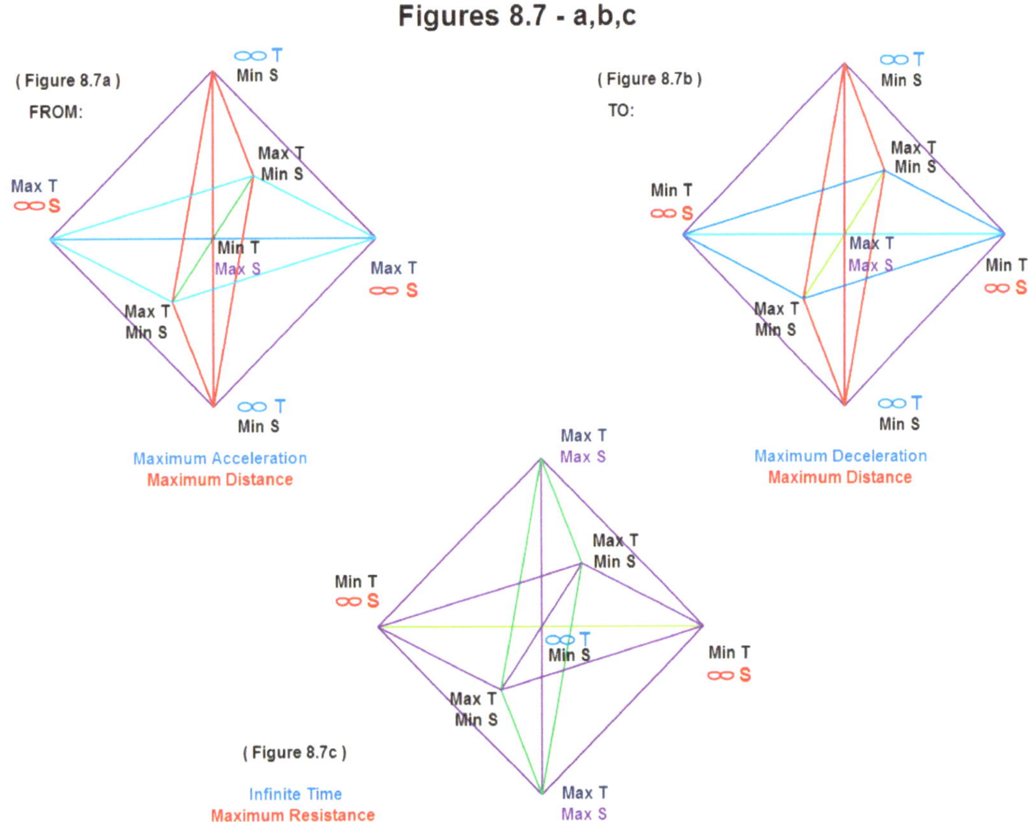

Where the prior six figures in the examination of energy to development, corresponding with the illustrated production of idealized progression towards evolutionary development via the immersive liquid inversion model, once again parody the induced perfunctory of mitotic cell division in a parallax of pertinacity, note also its emulsification into a condition of primary equilibrium emulous of the pictured molecule of hydrogen (see figure 8.8). Here, the achievement of a fifth dimensional relationship in relativity pursuant to the conceptual ideology of synergistic packing of energy prefigures a quality in existence that challenges imagination; although the achievement of a fifth dimensional sublimification, it is held in a geometric lathe of latent intrinsicalities personifying the integration of its interlacement in function: A sixth dimensional, idiomorphic informity based in the mathematical consignification of its concurrence.

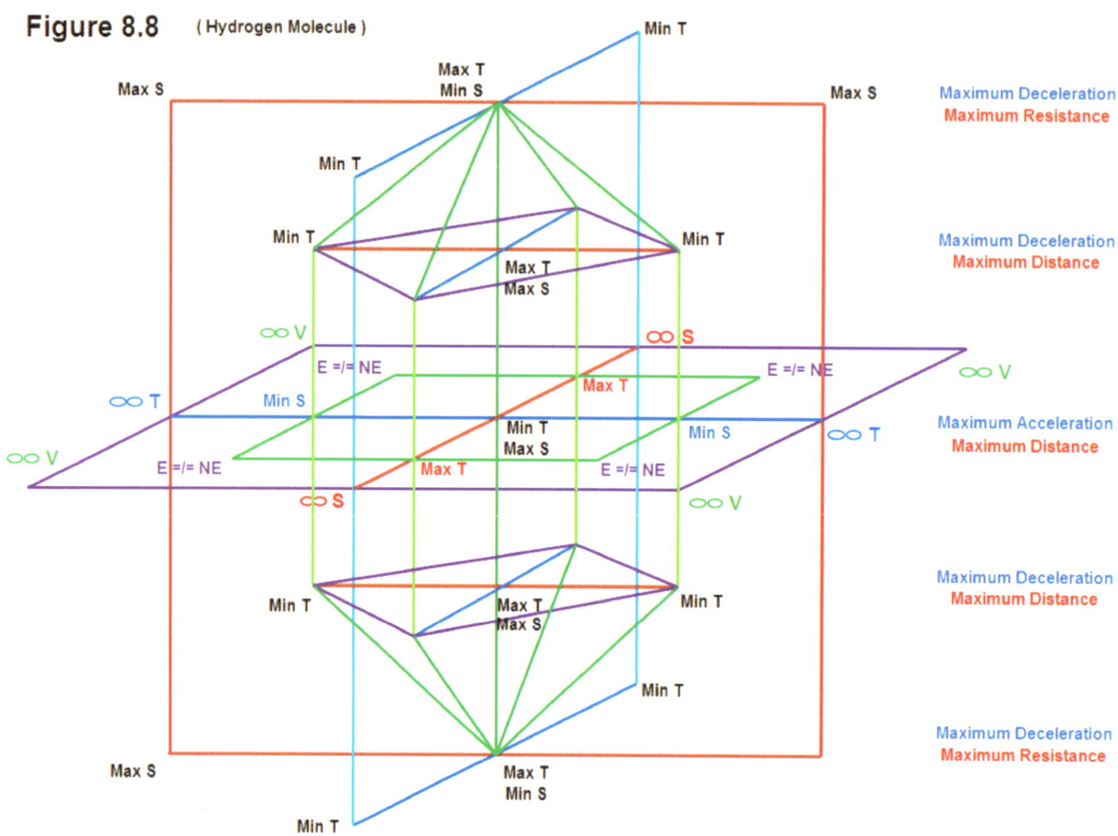

So, this pertinacious parallax in polarization of parity to pull pursues its own perpetuity in its percipience toward propagation (see figure 8.9).

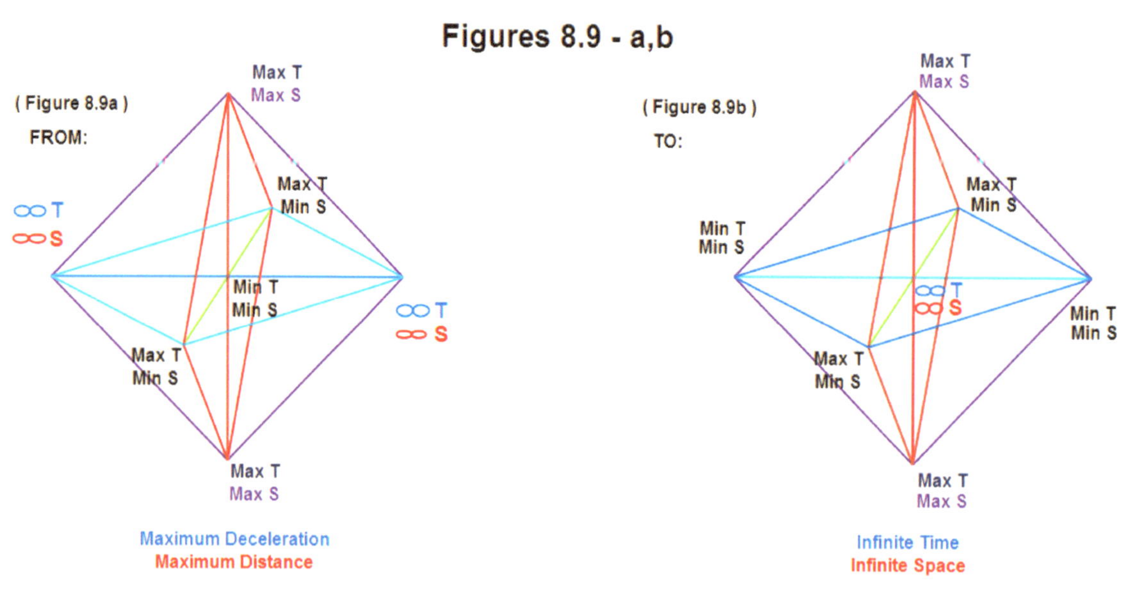

Closing Statements – Existence from Non-Existence

Such is it that an analysis of the development and evolution of energy and matter comes full circle in perspective to the four fundamentals of existence. Creation, i.e. the Universe, is a work in progress: Creation will never be completed as it is a product of its own definition - something from nothing (i.e. existence from non-existence), where nothing is unknown; A definition based in the dimensional parameters bounded by our senses. People need to define their own universe within the limitations of their own perceived physiological senses and their six sense of imagination (i.e. our ability to abstract ideas). Imagination is where faith comes into play because one must have faith in their own ability or someone else's ability to abstract ideas which are beyond their physiological senses.

Via reverse engineering, in part, I have chosen to model the Universe in dimensions of Space and Time, where Space and Time are the absolutes which define existence. Any measure of these absolutes requires reference points; i.e., one can not measure their position from just one point in Space (and Time), rather their position is the measure from a point in Space relative to another fixed reference point, other than itself. For without a "fixed" reference point, relevant to a pre-existing location, the measure between any two non-fixed points in Space would just continue to be dynamically redefined. The same logical representation occurs within events of Time as well, providing us with a union of these two differing dimensional abstraction along an axis of their relationship (or their dimensional juxtaposition) to each other. Since my model of my Universe requires both Space and Time for definition, the one can not really exist without the other. Indeed one without the other is non-existent to one's logical and even physiological senses. Therefore, by definition, the axis of their relationship becomes existence.

Yet where then is the other "fixed" reference point beyond this one-dimensional relationship? To measure existence we need a point from which there is no existence, i.e. a nonexistent point. In fact, our "fixed" reference point in more defined by its direction that by another position in Space and Time; by virtue of its ability to 'be', our position of existence is mirrored in direction from its ability to 'not be'. Existence achieves direction in difference, in "being". Although, envisaged as three separate directions - space, time, and velocity; Velocity is an entente of Space and Time that is an acknowledgement of their differences. Velocity is, veritably, that direction which is both directions, space and time, without being either direction. It can, therefore, be said that existence is achieved as velocity in velocity (i.e. acceleration) and that velocity is the measure of our existence.

Which brings me to a very important concept: non-existence, or that which can not be defined in the dimensional parameters bounded by one's physiological senses; Or more simply put, a concept outside the reference frame of Space and Time. A daunting task it is for one's sixth sense of imagination to qualify a quality of non-existence. I do not state that non-existence is not an any more "real" concept than existence, but that it does not have a state of 'being' within our reference frame of a Space-Time continuum. Of what form is this dimensional aspection of direction from existence to non-existence; a form seemingly without commencement or termination?

This measure of existence enlivens a constant second only to that of non-existence, and yet the first to be perceived. Non-existence, as the disintegration of Space-Time, becomes the opposite logical representation of events which occur within Space and Time, providing us with a disunion of these two differing dimensional abstraction along an axis of their relationship (or their dimensional juxtaposition) to each other: i.e., 'No Time' and 'No Space'. Bounded by the ideals of 'Infinite Space' and 'Infinite Time', the dimensional juxtaposition of 'No Time' and 'No Space' reveals itself as the "fixed" reference point from which existence is perceived.

Reproaching this idealism, even with two points, two separate and distinct ends to direction, to infinity (the recondite bridge between real and not real), a line would propose rather an altering state of velocity repudiating constancy. This expostulation of constancy abides in a reiteration of perception which could only be consistent to its change in direction from one point to another in recognition of an inflective state of perception in relativity. It must, in fact, stop to change direction toward each point. Rather the most expedient conveyance of such constancy for form, by shape, could not be a line which would remonstrate the ideal of constancy, but rather a circle: The affection of perpetuity which perceives its own identity in being. Oddly enough, if a circle of such a constant velocity were to be lain flat and viewed in this position, it would reveal the illusion of an altering state of velocity; A state congruent to an already imagined notion of velocity in perspective of a line. It is the circle that knows two directions while traveling one. It is the circle that can remain constant in form and travel from one non-existent point to another, purveying a parody of origin purporting an existence without beginning or end. This perception of direction, by form, from one point to another relates a cycle of 'being', via perception, where non-existence is that median of measure to converse directions for existence.

Therefore it is the actual structure of our Space-Time frame of reference that defines our relativity in existence. It is this relativity which defines our perception of 'being' based on the need for an apposable "fixed" point of reference in opposition to our nature of 'being'. It becomes the predication of getting 'something from nothing'.

Chapter 9

"The Distribution of Mass"

Prologue

As this process of conception and nurturing abets a similarly accommodating accrual, the design of this allegorical allusion to perpetuity in percipience toward propagation adumbrates a continuing mode of prognostication with which to predict prospective representatives in a concinnity of its primal precocity in energy. Note the expected evolutionary modification to the last illustration depicted in figure 8.9, eminently expressing the caricature of the helium atomic structure, based in the aspirations of animate engendering (see figure 9.1) demonstrated in the earlier presentment of the immersive liquid inversion progression (see figure 4.14 as presented in chapter 8), as previously presented from 'The Evolutioning of Creation: Volume 1'.

Figures 9.1 - a,b

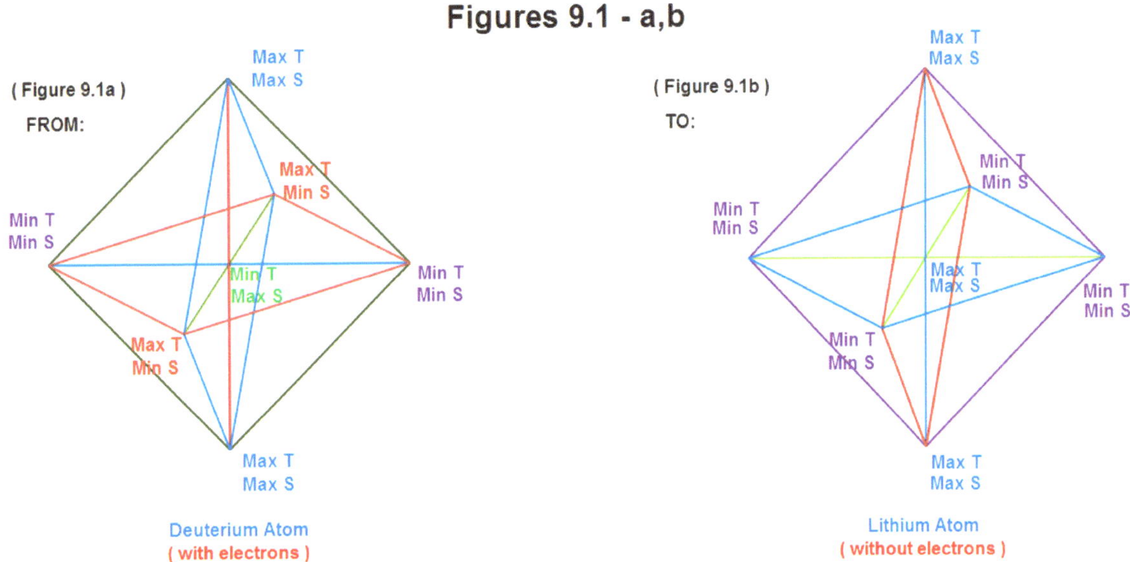

Again, as suggested earlier, the ability to include a conspicuity of electron distribution disappears because the acquired delineation with respect to its ideal can no longer undergo further adjustment in its devised establishment. I shall therefore backtrack a bit to include a renewal of a prior acceptation in perspective of the functional boundaries of energy that predisposes the relativity of existence; this, pursuant o its expected, projected expansion within the progression of idealized evolutionary development of synergistic packing in energy, infers a production in the distribution of mass corresponding to the preconcerted study in atomic orbital diagramming. And yet, to speculate upon a theory of development exclusive of an atomic evolution rather likens an elench (or sophism); for it can no express properly the significant expectation of polarization lying in anticipation of efficacious edification within the elaborated elicitation of the elucidated molecular enation (or material relationship).

Reminiscent to the parallel attitudes of Time and Space, as pictured in figure 8.5a, these fundamental forces are drawn, and can be held, in cohesion of poised polarity that personifies the estruation (or influence for strong impulse) of coitus. Indeed, a closer examination of the helium molecule expresses this primary polarization to the seemingly symmetrical design of structure: A symmetry of substantiation predicated to pervade the perfunctory perduration (or long continuance) of a universe peragrating (or passing through) a parody of suffusion in the parallelism of a matter and antimatter subsistence (see figure 9.2).

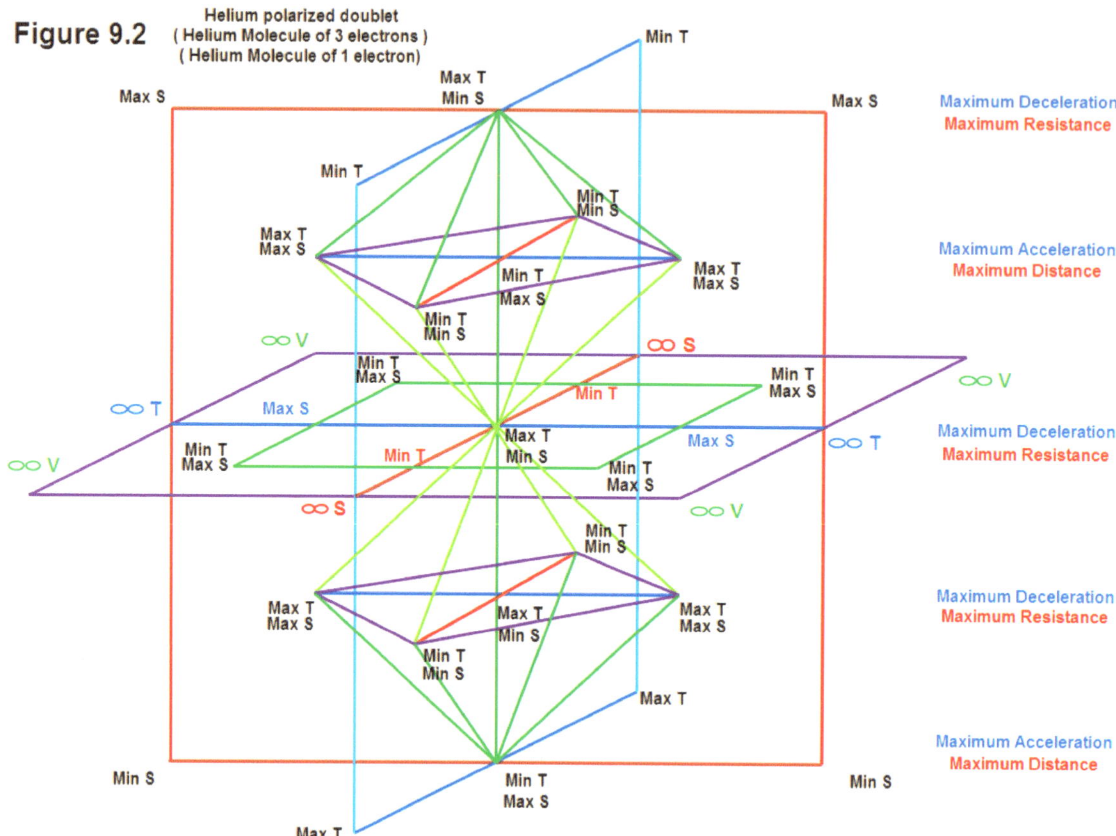

Figure 9.2 Helium polarized doublet (Helium Molecule of 3 electrons) (Helium Molecule of 1 electron)

Note the asymmetrical embodiment of a polarized doublet in the lower half of this depiction as the equanimity to a balance in equity of predominance in polarity. Still, this is not an expostulation of an antimatter existimation (or esteem), but rather of an antimatter equivocation, insomuch as it (antimatter) merely relates matter of anti-revolutionary motion with which to wrought an anti-charge among the gyrostatic interrelationships of the involved and intervolved particles. Here rather the notion of polarization examines the conditions which cause such molecular bonds to form adversely of a symmetrical aplomb (or poise, as in perpendicularity). Hence, one views a direction in velocity upon the apparent state of prevalence to the parallel attitudes of Space and Time in absence of an afferent (or the bringing into a central part) influence facilitated in the perpension of design alone.

This polarization, as demonstrative of the parallel attitudes in order and field of Time and Space viewed in planes, not only gives velocity a relativity in direction, but also examines a system of synchronous development in evolution alluded to earlier: A symmetrically ex parte excursion of velocity in existence based in the polaric (or polar) implications of involuted (or curled up in a spiral) motion that shapes its being upon the imbrications (or state of overlapping) of fundamental forces conforming within a dimension in function; which might also be loosely described as the mathematical interaction of being, as that which in itself is. The ambiguity of this description can be more aptly arbitrated from a quote by Bertrand Russel in which he states, "Mathematics may be defined as the subject in which we never know what we are talking about, nor whether what we are saying is true." Here, in the probability of a regenerative, molecular development to a relativity in deceleration via an articulated idiorepulsiveness (or state of self-repelling), there looms the possibility of an evolutionary gestation in the juncture of increasing motion in which the molecule might also be, but again, half of a larger concept in the superintendency (or oversight for the purpose of directing) of relativity. Because there is naught other than a recombination of this relevancy to the super-molecular (higher order molecular influence) sublimity (or state of being changed, condensed, or purified), does not discount or remonstrate regard to its rationality in existence.

In this way, the relative conciseness of demarcation unto the distribution and combination of mass, proliferated in the relegation of non-existence, protracts the inherent proclivity of a transition to energy. One need only imagine upon what rapport, in rapt rapture of existence, would such a development know of its remuneration o a birth in the greatest of all supernovas, the 'Big Bang' if you will, preceding its conception. One might even adumbrate (or foreshadow something) upon the design of such augmentative concurrence of energy unto the evolution of the inherent helix heritage heralded in its progenitive profundity (or abyss, as reference to non-existence) only as 'DNA' (deoxyribonucleic acid – the super-molecular structure occurring as the basic structure of all living matter). It is then that such a review might recognize a substantiation of superintendency to a superstructure of evolutionary development (see figure 9.3).

Figure 9.3

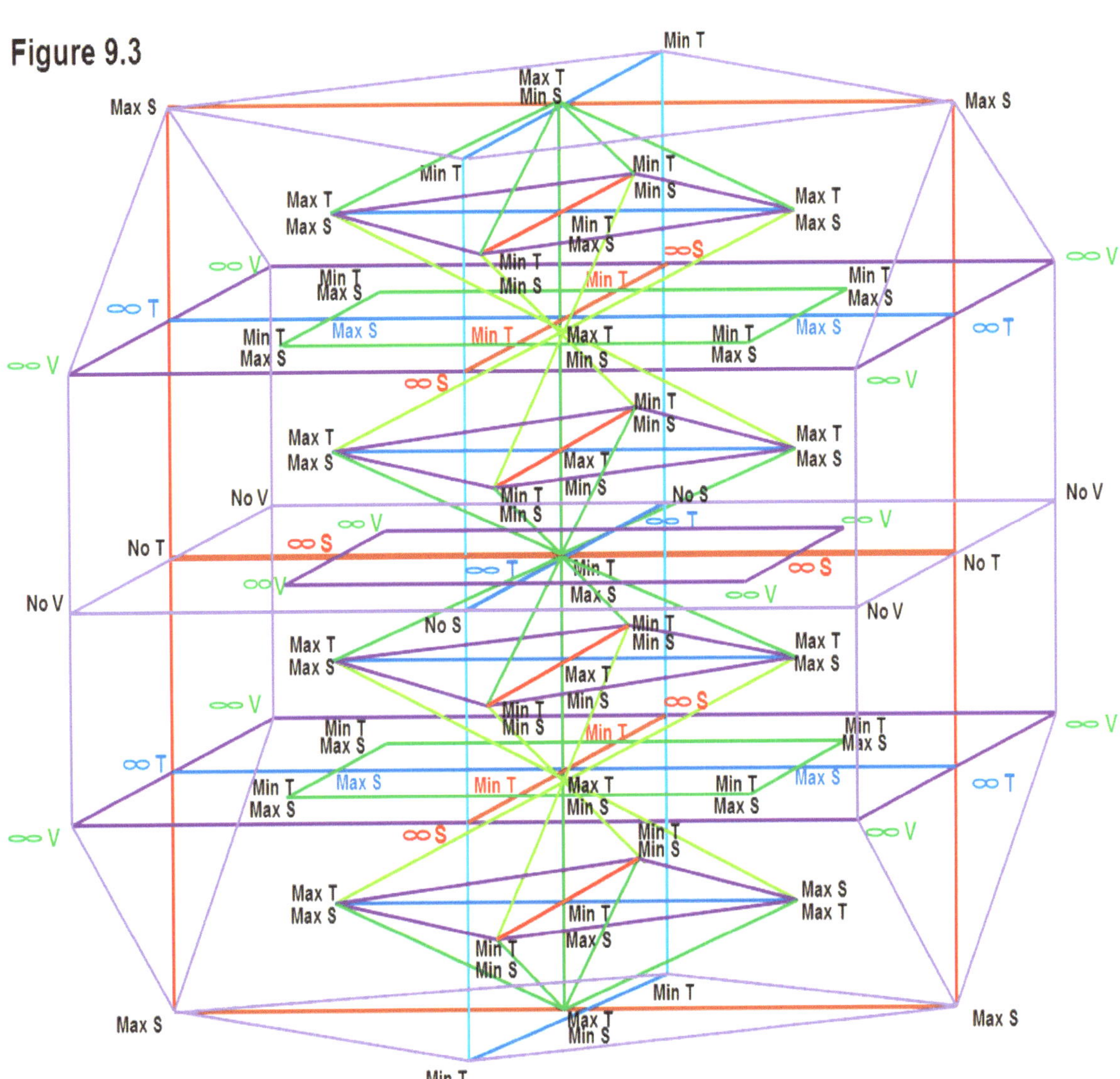

It reestablishes the afferent aplomb of influence to a continuum based in all six dimensions in an understanding to the assimilated processes of cell division and immersive inversion as well as provides an accent of affinity advocating affection for a' fortiori (or a conclusion that follows with even greater significance) than an afflatus (or inspiration) for aesthetics or proliferation.

An Evolutionary Pattern of Development

Such a gregarious parallel in this institutive evolutionary development of creation might prove to be the inspirational evocation of entropy in inertia whereupon matter builds in the atrophy of velocity. Certainly, it would seem to suggest the type of subsistence necessary to nurture energy to a substantiation of sublime resistance as revered of the ultimate supernova, the 'Big Bang'. And still, this is not the only pattern of such subsistence with which to verify this rationality; upon a liberal examination of the periodic law to elemental production, one views a probable repetitive progression to a subsequent inverse relevance to the relationship of relativity now existing, a pre-existing impetus incubating the indefeasibility of inspiration. Where electron (e), neutron or neutron forming (N), positron (p) and proton or proton forming (P) are utilized in the analysis of this pattern below:

EVOLUTIONARY ELECTRON PACKING	ATOMIC EQUIVALENCE	
e	1e,0N,0p,0P	(electron)
e p e	0e,1N,0p,0P	(neutron)
e pep e	2e,0N,0p,1P	HYDROGEN
epe pep e	1e,1N,0p,1P	DEUTERIUM
epe pep epe	0e,2N,0p,1P	TRITIUM
e pep e pep e	3e,0N,0p,2P	HELIUM-2iso
e pep epe pep e	2e,1N,0p,2P	HELIUM-3iso
epe pep e pep epe	1e,2N,0p,2P	HELIUM
e pep e pep e pep e	4e,0N,0p,3P	LITHIUM-iso
epe pep epe pep epe	0e,3N,0p,2P	HELIUM-iso
e pep e pep epe pep e	3e,1N,0p,3P	LITHIUM-4iso
epe pep e pep e pep epe	2e,2N,0p,3P	LITHIUM-5iso
e pep e pep e pep e pep e	5e,0N,0p,4P	BERYLLIUM-iso
e pep e pep epe pep e pep e	4e,1N,0p,4P	BERYLLIUM-iso
epe pep epe pep e pep epe	1e,3N,0p,3P	LITHIUM
epe pep epe pep epe pep epe	0e,4N,0p,3P	BERYLLIUM-iso
E	1E	(electron)
E N E	2E,1N	(unknown)
EN P NE	2E,2N,1P	TRITIUM

E N P E P N E	3E,2N,2P	HELIUM
E N P E N E P N E	4E,3N,2P	HELIUM-iso
E N P E N P N E P N E	4E,4N,3P	LITHIUM
E N P E N P E P N E P N E	5E,4N,4P	BERYLLIUM-iso
E N P E N P E N E P N E P N E	6E,5N,4P	BERYLLIUM
E N P E N P E N P N E P N E P N E	6E,6N,5P	BORON
E N P E N P E N P E P N E P N E P N E	7E,6N,6P	CARBON

It is the aftermath of this 'Big Bang' that changed the very order of relativity into the disjointed physics we are left with to trace and reconstruct our own etiology. Such is it that this relativity facilitates a view of comparable progression, lending rationality to its relevancy. Where electron (E), neutron (N), proton (P), and an unknown recombinant malconformation (or disproportion of parts) of atomic structure (X) are utilized in the analysis of this pattern below:

ATOMIC EVOLUTIONARY PATTERN	ATOMIC EQUIVALENCE	
E	1E,0N ,0P	(electron)
N	0E,1N ,0P	(neutron)
P	0E,0N,1P	(proton)
E **X** P	1E,0N,1P	HYDROGEN-1
E N **X** P E	2E,1N,1P	DEUTERIUM
E N E **X** P N E	3E,2N,1P	TRITIUM
E N E N **X** P E N E	4E,3N,1P	HYDROGEN-4
E N E N E **X** P N E N E	5E,4N,1P	HYDROGEN-5
E N E N E N **X** P E N E N E	6E,5N,1P	HYDROGEN-6
E N E N E N E **X** P N E N E N E	7E,6N,1P	HYDROGEN-7
E P **X** P E	2E,0N,2P	HELIUM-2
E P E **X** N P E	3E,1N,2P	HELIUM-3
E N P **X** P N E	2E,2N,2P	HELIUM-1
E N E P **X** P E N E	4E,2N,2P	HELIUM-4

E N E P E **X** N P E N E	5E,3N,2P	HELIUM-5
E N E N E P **X** P E N E N E	6E,4N,2P	HELIUM-6
E P E N **X** P E P E	4E,1N,3P	LITHIUM-4
E N E P E **X** P P E N E	5E,2N,3P	LITHIUM-5
E N E P N **X** P P E N E	6E,3N,3P	LITHIUM-6
E N P N E P E N P N E	4E,4N,3P	LITHIUM
E N E N E P E **X** P P E N E N E	7E,4N,3P	LITHIUM-7
E N E N E P E N **X** P E P E N E N E	8E,5N,3P	LITHIUM-8
E P E P E **X** N P E P E	5E,1N,4P	BERYLLIUM-5
E P E N P E **X** E P N E P E	6E,2N,4P	BERYLLIUM-6
E P E N E P E **X** N P E N E P E	7E,3N,4P	BERYLLIUM-7
E N E P E N P E **X** E P N E P E N E	8E,4N,4P	BERYLLIUM-8
E N P E N P N P N E P N E	4E,5N,4P	BERYLLIUM
E N E P E N E P E **X** N P E N E P E N E	9E,5N,4P	BERYLLIUM-9
E N E N E P E N P E **X** E P N E P E N E N E	10E,6N,4P	BERYLLIUM-10
E N P E N P N E P E N P N E P N E	6E,6N,5P	BORON
E N P E N P E N P **X** P N E P N E P N E	6E,6N,6P	CARBON

Thus a new order arises from the old, as I retire this speculation as a reversionary analysis of the former study in the evolutionary gestation of an energy-like development in existence, to which I shall continue where I left off with figure 9.2 as the protuberance of propagation in propensity of progression. In adopting an expanded configuration to the design of this developing resistance in velocity towards increasing mass within a perpension of its empirical relation to the orbital diagramming of its atomic structure, I believe I can further demonstrate this atomic propagation of energy which seeks to abrogate a sentience of non-existence. As this graphic characterization continues, it too shall lose the perspective of its electron relationships upon the imposed restrictions to the limits of its design. Note the continuing transmutation of developing energy to the tenable transitions of demarcation. A demarcation transfixed in the infirmity of an 'inflective paralipomena' (an altering supplement containing things omitted) in syntax of its recombinant propriety to a nature of temporal polarization (see figures 9.4 and 9.9). Such are the transgressions of temporization which seek to trisect the tremulous transcendentalism of existence in the remuneration of energy into mass.

Figure 9.4

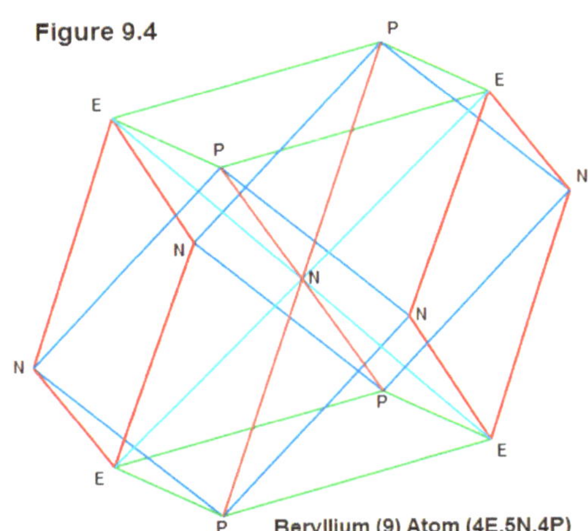

Beryllium (9) Atom (4E,5N,4P)

Figure 9.5

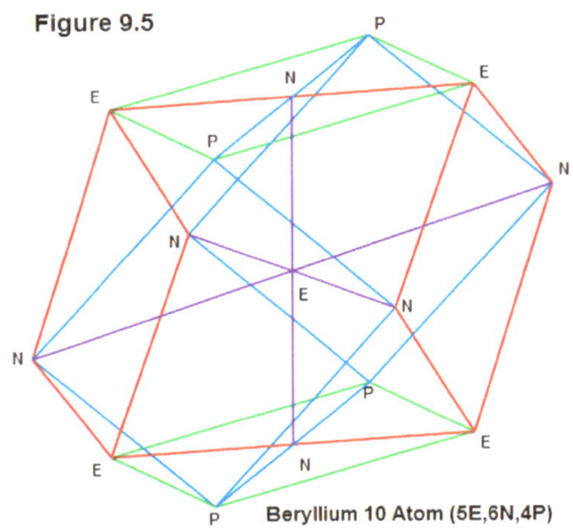

Beryllium 10 Atom (5E,6N,4P)

Figure 9.6

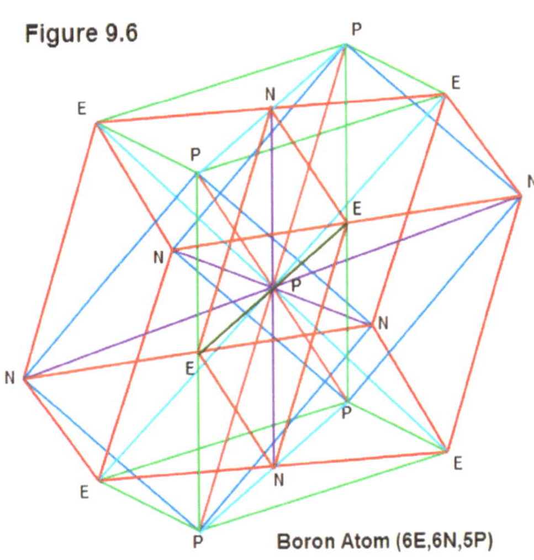

Boron Atom (6E,6N,5P)

Figure 9.7

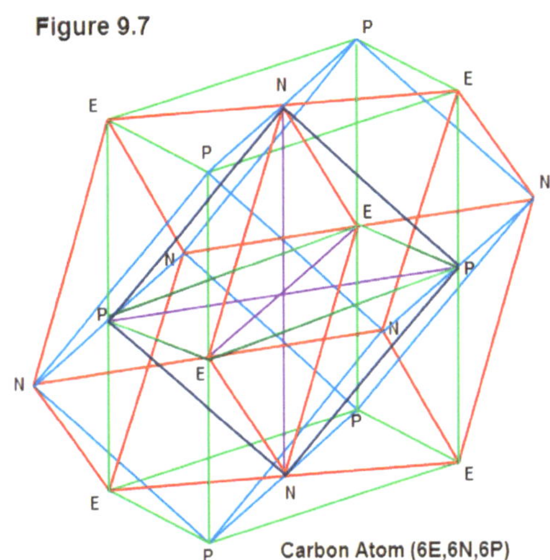

Carbon Atom (6E,6N,6P)

Carbon accounts for the aspection of an unrealized neutron in the formation of an additional proton within the stabilized structure of a more primer, helium-like core; liken to the seeding of the atom by design, it is a configurable transition from Boron which viviparously promotes a vitality of essence in the supported structure of the helium atom (as seen in figures 9.6 and 9.7). Yet it is the more polar-like transition for the simultaneous formation of a proton and a neutron in the concurrent interluency (a flowing between) of atomic structure (as seen in figures 9.8 and 9.9) that intriguingly mimics the interphase of cellular mitosis. It is an acknowledgement of a more polar intermediary, or incorporeally polarized force, which was previously ensigned to an earlier comparable progression of relativity for the creation of the hydrogen atom. Each succeeding formation can then be imagined to imitate a preceding, less complex configuration, as in the lethargic engendering of an 'enforced energesis' (a liberating of energy into simpler composition), if you will; the exigency to exflect (the critical need to cause to turn outward) an exestuation (an agitation by heat) in resistance.

Figure 9.8

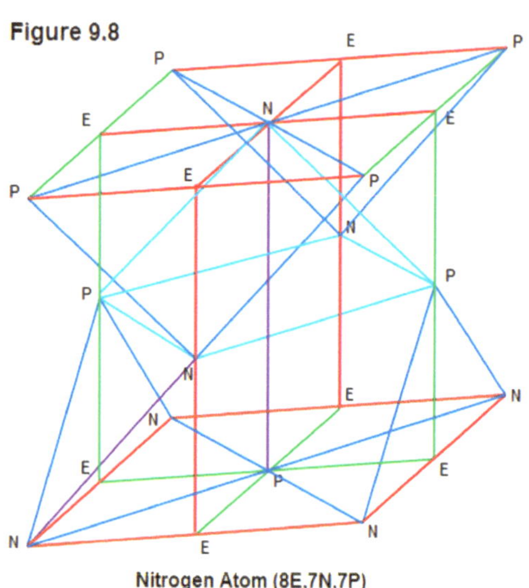

Nitrogen Atom (8E,7N,7P)

Figure 9.9

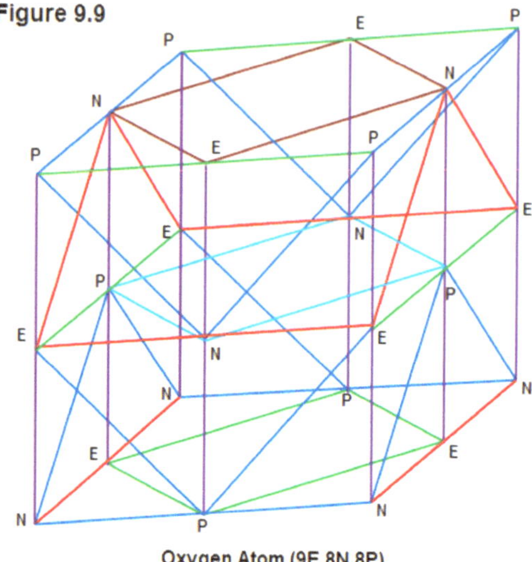

Oxygen Atom (9E,8N,8P)

Note the recombination of this atomic structure for oxygen in figure 9.9 to this expectant polar-like influence brought about by either that speculation of a larger purview in relativity of atomic forces or some innate speculation procured via its acquisition for an accommodating acculturation. The key here is the polarization of nitrogen in figure 9.8, further verifying the preceding concept of 'attitudinizing' a parallel expectant, namely Oxygen, from a perpendicular involution, namely carbon. This homogeneous parallelism is recognized as the idealization to a position of cellular prophase, to what can be attributed to the emulation of its emulsification assimilated in mitotic cell division. The Oxygen atom takes on the polarized vivacity of velocity in the paralleling of the proton and neutron planes in a configuration which viviparously promotes a vitality of essence in the supported structure of the helium-like molecule; liken to the seeding of the molecular condition by design.

Note the perfectly completed 'metaphase' condition of the indivisibly individualized character of the Neon atom in perpendicular design to the focalization of velocity in resistance; a design which would seek to induce that incipiently enceinte quality of energy-like development to incite procreation (see figure 9.10 and 9.11).

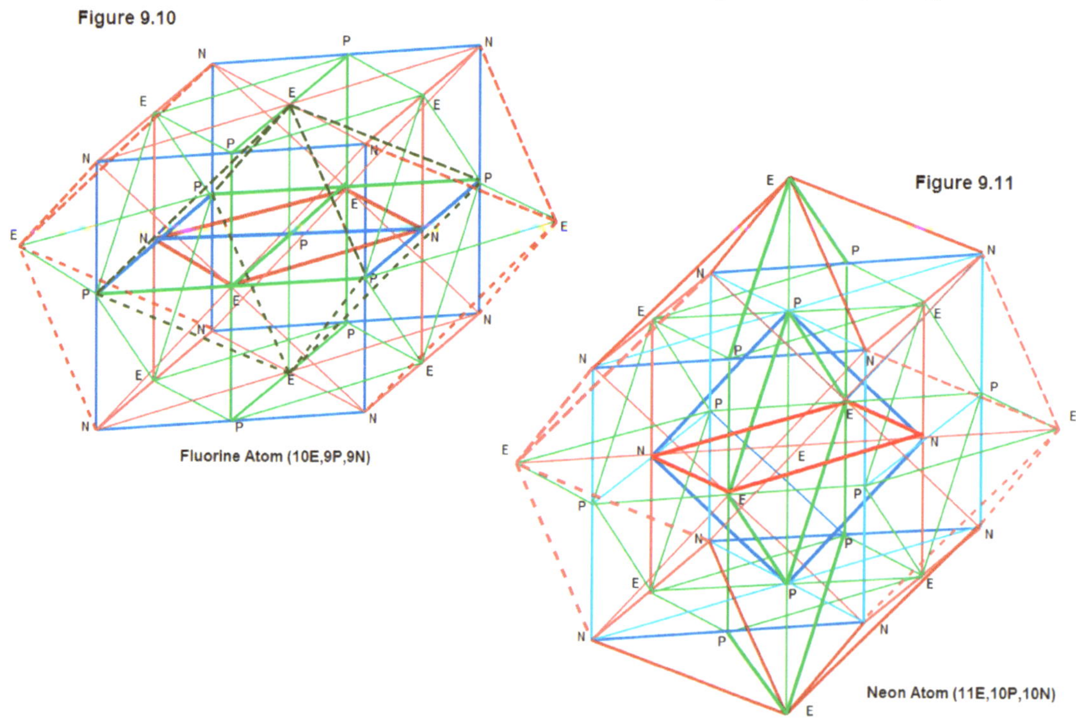

Figure 9.10

Fluorine Atom (10E,9P,9N)

Figure 9.11

Neon Atom (11E,10P,10N)

Such is it that this quality set up a repetitive progression in which one should view the reformation of the Lithium atom within the construction of the Sodium atom, and on through to the Aluminum atom (see figure 9.12 thru 9.14).

Figure 9.12a

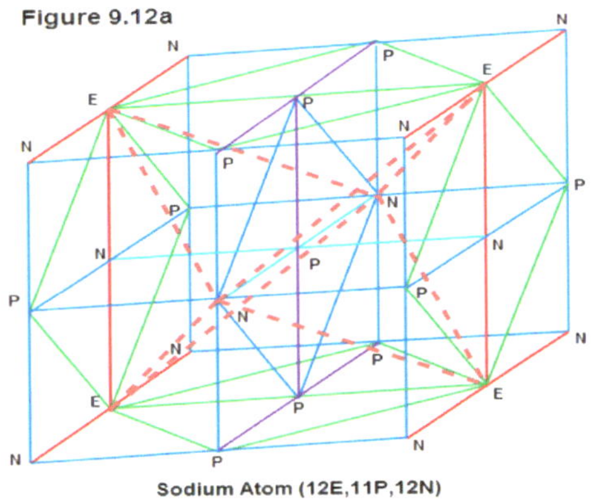

Sodium Atom (12E,11P,12N)

Figure 9.12b

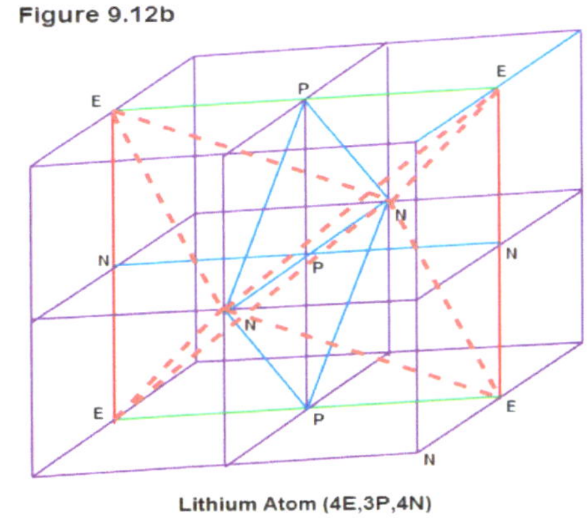

Lithium Atom (4E,3P,4N)

Figure 9.13

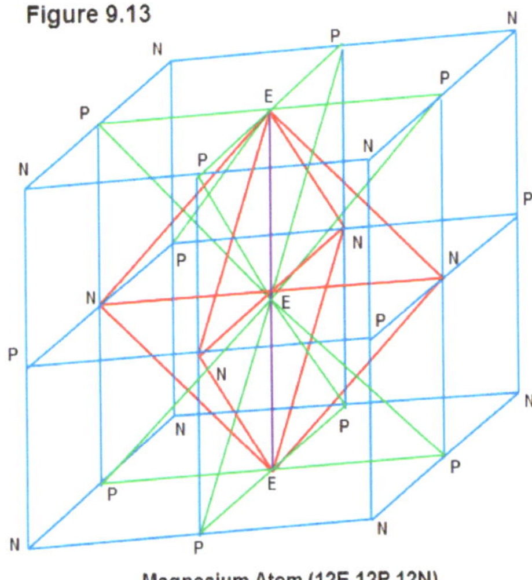

Magnesium Atom (12E,12P,12N)

Figure 9.14

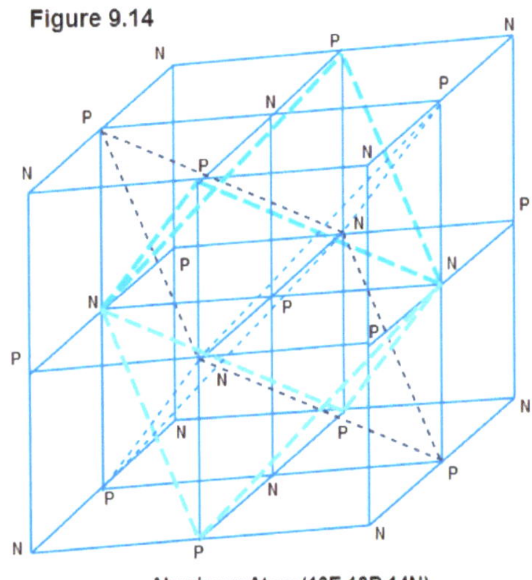

Aluminum Atom (13E,13P,14N)

This repetitive progression bears the verisimilitude of re-verification in recognition of other such anatomically atomic analogies. Note the design of figure 9.5 repeated within the horizontal planes of the Magnesium atom, depicted in figure 9.13, while retaining the vertical bordering planes of a precedent purview found in figure 9.4.

The Aluminum's atomic configuration conspicuously admixes this precedent purview in all encompassing concrescence planes engendering the central vertical posture of the boron's proton structure, in both vertical and horizontal attitudes, within the concurrence of a predictable, six character, neutron inner focus (see figures 9.14 and 9.6). Looking back, one should now be able to predispose a predilection of the helium's atomic condition within the Neon's atomic structure (see figures 8.3e and 9.11). Additionally, one can envision a commingling of figure 9.5's vertical plane concerted to the central horizontal level of the Neon's delineated representation. Leaving on to expect the prevalent precocity of the Hydrogen atom to appear at the foci of the Fluorine's atomic structure upon a de-evolution of a professed repetitive progression.

This organization, built upon a foundation of energy-like precocity toward a more progressive procreation in the relativity of evolution, would seem to characterize a system of prime orders developing as the prototypes of building blocks to be used within the make-up of the universe. This system of primes, therefore, insinuates that all energy and/or matter is rather that recombination of parts, heretofore created unto the very dawn of existence, forged into shapes, either by fission or fusion or other imagined conglomeration of resistance, which defines their relationships among one another in varying degrees of decelerated subsistence.

Figure 9.15

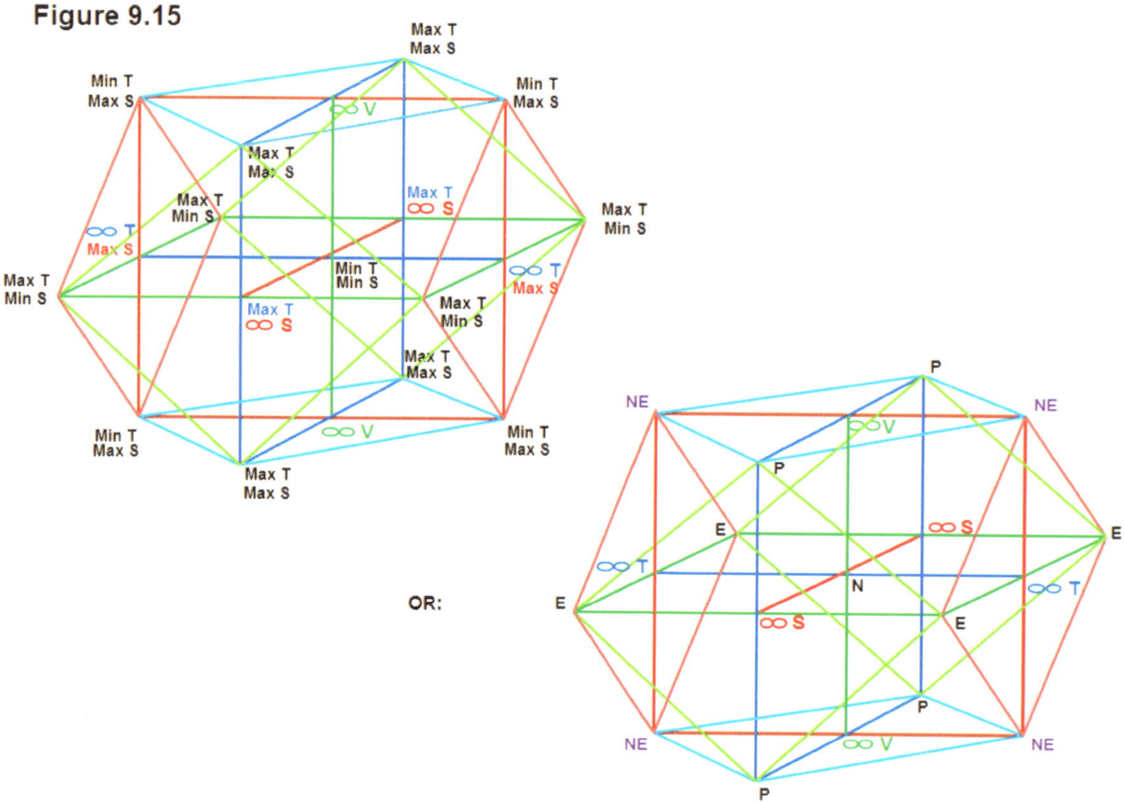

Such is it that it can therefore be theorized, as reminiscent of a speculation unto the sought after paragon of a larger purview in the superintendency of its relativity expressing that representation of inflection to maximum velocity portraying an invigoration of conception toward evolution, that existence upon inception reflect a fixity of the four fundamental forces of energy with the fusion of four protons in the prepotent intropression (the influenced internal pressure) of fission within the 'idiorepulsive schizogenesis' (self-repelling reproduction by fission) of electron impedance as the *'in situ'* insistence inspiring an intumesce of irruption *'in vacuo'*: The invocation of neutron intinction (see figure 9.15).

Closing Statements – Forces of Motive Matter

Yet, such organization is more than a caricature of relativity within the confines of its self-made continuums based on the involutions of velocity. Varied via resistance and direction, as studied of the immersive inversion and mitotic processes, the inference of a polaric influence to the innate, essential character of motion and energy qualifies a sequence of nurturing to a synergism of persistence known as current. Therein these processes enliven a path of least resistance in their affinity for potential predominance of velocity toward acceleration. One can more easily visualize this polaric influence, in design of these forged relationships, among the efficacious educations of isomeric compounds, where shape is prioritized over content. More importantly, this path of least resistance (current) further seeks to suggest the feasibility of an evolution in energy as the primary constructor of form that projects a model emphasizing the indivisibility of Time and Space as the boundaries of existence; An existence to which motion is the measure of directions to a focus realizing velocity as the contention of vortex and centrifugal forces.

Understanding this focus, one acknowledges its relationship to a practical illustration of prior analysis in temporal refraction: The greater the acceleration, or the lesser the motion (or perception of motion), the greater its inertial frame of reference; the greater the deceleration, or the greater the motion (or perception of motion), the lesser its inertial frame of reference. More simply stated, the faster one's speed, the slower he (perceives that he) goes as in the greater his distance (or as in his enhanced ability to perceive greater distance) there is to travel. This is seemingly a contradiction of empiricism and takes on a more favorable acceptance when considering the notion of an eighth or even tenth dimensional existence, acknowledging a limited continuum in Time and Space, of speed of light or greater, whereby velocity loses direction to a measure of Minimum Time and Maximum Space in an inflection (a bounding) of infinity; where infinity imbibes a perception of non-existence.

It is this concept of current, therefore, that scribes its impression of form into an expression of shape toward that comprehension of existence as that sphere of perception in measure of converse directions to expansion and contraction. Thus, for current to exist, one must merely prove motion of/in polarity via shape: That is 'an affinity for' and/or 'a repulsion to' a direction in velocity, in recognition of the true relativity of perception. This conversion of dimensional ambivalence contorted from an end point of maximums to a curve of contumacy, noting the convexity of its origin, was first infused in that illusion of imagined travel in a velocity of increasing deceleration, increasing motion (or perception of motion), in which a quality of parity portends its own equanimity in the limits of its boundaries to existence.

Such is it an added aspect of velocity that current sublimate in personification of peripatetic periphery congruent to that notion of superconductivity: The ability to accelerate a body of conductivity providing for a conflux in continuity of conformity; perpetual motion. Here the phase difference of the induced magneto-motive force and magneto-resistance produces a relationship that maintains the inertial stability of current, in deceleration of energy to mass, along magnetic tubes of flux prepossessing a prepotency of magnetic hysteresis upon the rippled effectuation of its flux density. One might perfect a model of this ripple effect in sublimation of an electromagnetic field unto a divining of energy to mass relative to its evolutionary gestation, similarly as proposed in figure 9.16 (see figure 9.16).

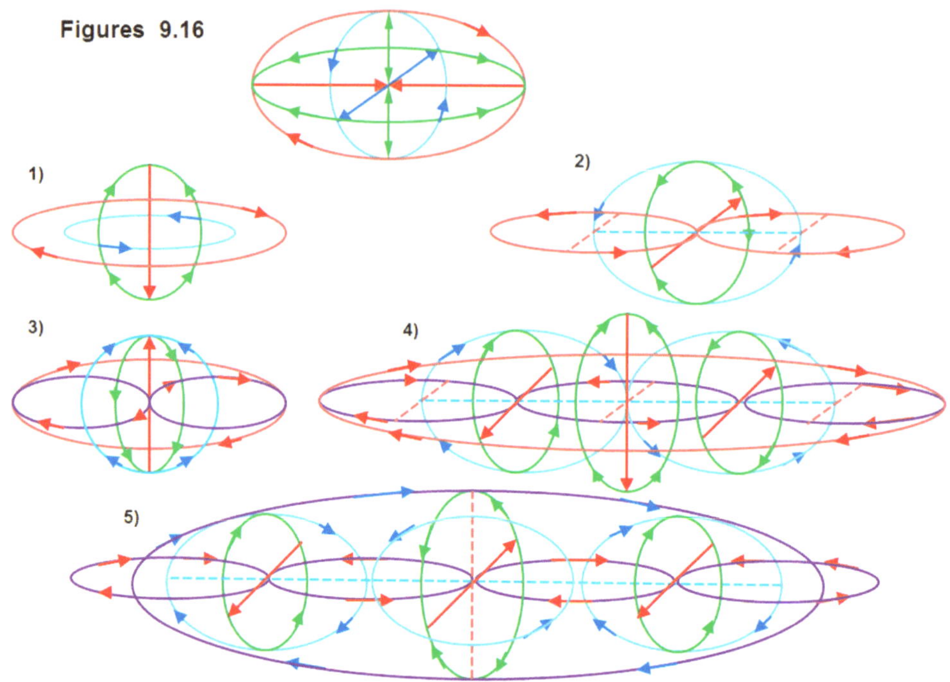

Here one notes the fission of Time upon the simultaneous fusion of Space to which relativity maintains equilibrium in the balance of that expression of velocity imbibing the infusion of percipience in existence. Here, the negativity of identity to being is expanded upon its own reflection in repulsion to an origin of non-existence, while its positive counterpart is forced into an affinity for that inflection of infinity to an origin of existence wherein it maintains its culmination for being (its expression 'to be'). A similar process of neutron disintegration envisions this concept in the high speed ejection of an electron, and subsequent collapse of its electron field, for the creation of a positron and its positron field. Eventually this repulsion takes on that character of displacements in existence whereby resistance ('all Time' having been displaced in motion as motion is displaced in resistance) is then displaced in distance ('all Space'), realizing an affinity to its complement. An affinity that upon return is permitted into its reciprocal field of origin, but again it remains held in its 'repulsion to origin' due to the polaric imbalance induced upon its being displaced; an imbalance known similarly as a super magnetic monopole. This mono-polar imbrication (overlaying layers) of identism, representing the enceinte (pregnant-like) quality within the epitome of nature for an evolution seeking equanimity, then remedially divides upon a reluctance to its relevancy in such magnetostriction (or magnostriction). It is this reluctance to magnetostriction that embraces a field of instability; an instability inspiriting an influx of kinetic force to its potential exertion.

Imbuing this compulsive complacency for compromise, within an evaluation of its conformity, energy exists as a form of matter in which a kinetic state, as the impassioned vestige of the inherent vicissitude in evolution, 'impregnates' a potential state for positive enthalpy and negative entropy. This evaluation, although an allegorical conjecture to the relativity of a 'being' in existence to be sure, adumbrates an allusive ambiguity belied upon too fastidious a propensity for energy. This uniquely remarkable genesis of energy circulation, via its transmission within the substantiation of being in existence, represents a focal point of force constituting a natural gravitation (or directional movement) in the electromagnetic alignment of primordial permutations to an 'evolutionary probation', if you will. Such becomes the primal stages of a perceptibility to the random forces within and without its medium. This electromagnetic perceptibility ingratiates a register of existence in its nature of 'being', culminating a precocious precipitation in the composite predisposition of its nature in relativity, in the random emulsion of its confluent consolidation. Such a consolidation could be the postulate of inherency in the proliferation intended as the intrinsic intensification of physiognomy, the continuous blueprint to the contingent make-up of a semblance in the resonance of a reminiscence in perception, known also as the 'DNA of evolutionary rejuvenescence' in organic ordination.

Chapter 10

"An Exegesis of Evolution"

Prologue

Although I have excogitated upon the expediency of an evolutionary process in existence, I feel it is necessary to discuss the nature of evolution against what I deemed to be previous misconceptions or prefabrications. Evolution is represented as the somewhat mystical intervention within the instability of 'being' which exists now as it has existed from inception and as it will exist unto termination. This verity is proven upon each living entity's ability to influence its own revivification unto their offspring; a revival which in itself regulates a gesticulation of modification in the gestation of such progeny. Therein evolution is enunciated in the periphery of our perceptivity, and presents no threat whatsoever to a philosophy of creation. Rather, what it does is to merely approach an objectivity in the optimism of an omnipotent occlusion: The constant of relativity. This constant of relativity in existence knowing 'being' in evolution moreover portends a comprehension of an origin in creation as the only irrefutable postulate to an operative infinitude. As the enigma of a truly one dimensional singularity, all exists as the product of one, or oneness, in omneity.

Once again the 'Bible' is remanded upon that familiar parallax, "Understanding from truth", as the oldest rendition of evolution, conceived over four thousand years ago. In it, creation was noted to follow along such lines of reasoning as ratiocinated in the comparable ideology of Darwinism, most publicized of these 'free' thinkers. Here within the 'Book of Genesis', one of our first and alas unknown free thinkers tinkers with the art of discovery unto the very roots of nature, its fundamental formation, and its precedent conception. Here is recognized a spherical insistence to an environment of man as the "dome called sky" separated the waters above from the waters below, under the "appearance of light": An atmosphere of evaporative partibility. Here is recognized the regenerative respites of organic respectability to an invigoration of particularity in parasitism as the "living creatures" to the previously mentioned "vegetation". Here is noted a "making of man", both male and female, in the image of all things, or "our image"; 'our image' relating to all things pre-existing: Plants, animals, and God alike. Such is it that the 'Bible' relates one of the first theories of evolution by interfusing a logical science-like rendering in theosophy, if not theology.

Understanding the term 'God', as origin for 'being' toward the infinite imperceptions of velocity, one takes the first step toward deciphering the universal physiognomy of theism as the scientific principles of Present-Time analysis based in Einstein' 'Theory of Relativity'. The ability 'to be' before 'it is', suggests this relationship of Present-Time analysis as that which one can only be imagined as non-existent. Thus, one begins to realize that each evolutionary modification is subjected to the interpretation of 'being created' via an infusion of non-existence in existence: e.g., via the 'Bible', "God ... blew into ... the breath of life, and so ... became a living being". Creation, origin, becomes the speed of perception that begets inception: i.e., that which 'has already been' is known to be before it is; omniscience. It is description not unlike that of an expectant potential insistence as foreshadowing a kinetic persistence. Creation, as that factor of probability seeking to preclude an evolution, rather predetermines the nature of evolution as a reflection of its own inner character and functional 'being'.

Evolutionary Development in Motive Matter

Evolution is that process which exists in the furthered complexity of organization, in matter, through a random occurrence building upon and based in necessity and past survival. Such a selection can be traced in the rapid process of gestation, as discussed earlier, where it follows its pattern for being upon its varying stages of development. Note that the pattern of any random mix is all those selections incurred in order from origin to completion, providing for a permanent record in the evolution of its being. The foundation of all organic life exists is just such a random selection and accretion of elements toward achieving a polar-like balance between its inherent existence from non-existence. While many look for dramatic demonstrations of an evolutionary process, there remains the simple underlying truth of variations formulated and maintained from one generation of being to the next. To deny evolution is to deny an existence from within the origin of all life, sometimes termed as 'God'. Yet the terminology is not as important as an understanding of its meaning and relevancy to 'being', existence, and subsequent persistence toward individualizing life force.

By virtue of these randomly formulated relationships based in early energy formation and their past survival selections maintained via early energy circulation, these primes endure in the predesignations of preference called evolution. Man is still, even now, unraveling this enigmatic relationship. The four primes extricated to exist in all organic life on Earth are hydrogen, carbon, nitrogen, and oxygen; with emphasis on carbon as the most essential component. Curiously enough, the latter three reciprocals, heretofore relating a shifting of, or recombination in, the order of atomic development from a perpendicular involution to a parallel expectant (as noted of a polarization in energy toward current), also appear together in order of energy relationships within the periodic table of elements. This relationship remains as the pattern of random mix retained and inherent in all evolution of organic life. It is a prime order patterning of evolution that maintains the universal structure of our cosmography.

Of these four primes, hydrogen, carbon, nitrogen, and oxygen have formulated a relationship to provide for the circulation of fluids within all organic life on Earth that mimics the early energy circulation. It is a past survival selection that is used to retain the ability to uniquely individualize organisms. Examining such a familial pattern of development as demonstrative of both plant and animal (or organic) energy and elemental cycles, one notes a conglomeration toward an inner density bounded by four pyrrole nitrogen atoms bonded to four hydrogen-carbon molecules noting a striking polaric amalgamation of oxygen to one side of this unit.

Pyrrole, as explained via the 'Encyclopedia Britannica' is:

"... any of a class of organic compounds of the heterocyclic series characterized by a ring structure composed of four carbon atoms and one nitrogen atom. The simplest member of the pyrrole family is pyrrole itself, a compound with molecular formula C_4H_5N. The pyrrole ring system is present in the amino acids proline and hydroxyproline; and in coloured natural products, such as chlorophyll, heme (a part of hemoglobin), and the bile pigments. Pyrrole compounds also are found among the alkaloids, a large class of alkaline organic nitrogen compounds produced by plants.

In heme and chlorophyll, four pyrrole rings are joined in a larger ring system known as porphyrin. The bile pigments are formed by decomposition of the porphyrin ring and contain a chain of four pyrrole rings."

Porphyrin, as explained via the 'Encyclopedia Britannica':

"... any of a class of water-soluble, nitrogenous biological pigments (biochromes), derivatives of which include the hemoproteins (porphyrins combined with metals and protein). Examples of hemoproteins are the green, photosynthetic chlorophylls of higher plants; the hemoglobins in the blood of many animals; the cytochromes, enzymes that occur in minute quantities in most cells and are involved in oxidative processes; and catalase, also a widely distributed enzyme that accelerates the breakdown of hydrogen peroxide.

Porphyrins have complex cyclic structures. All porphyrin compounds absorb light intensely at or close to 410 nanometres."

I am, of course, referring to the life's blood mechanism of Earth's two predominant organic life forms; the basic molecular structure of porphyrin rings for chlorophyll and hemoglobin noting twelve specific sites for classification (see figure 10.1).

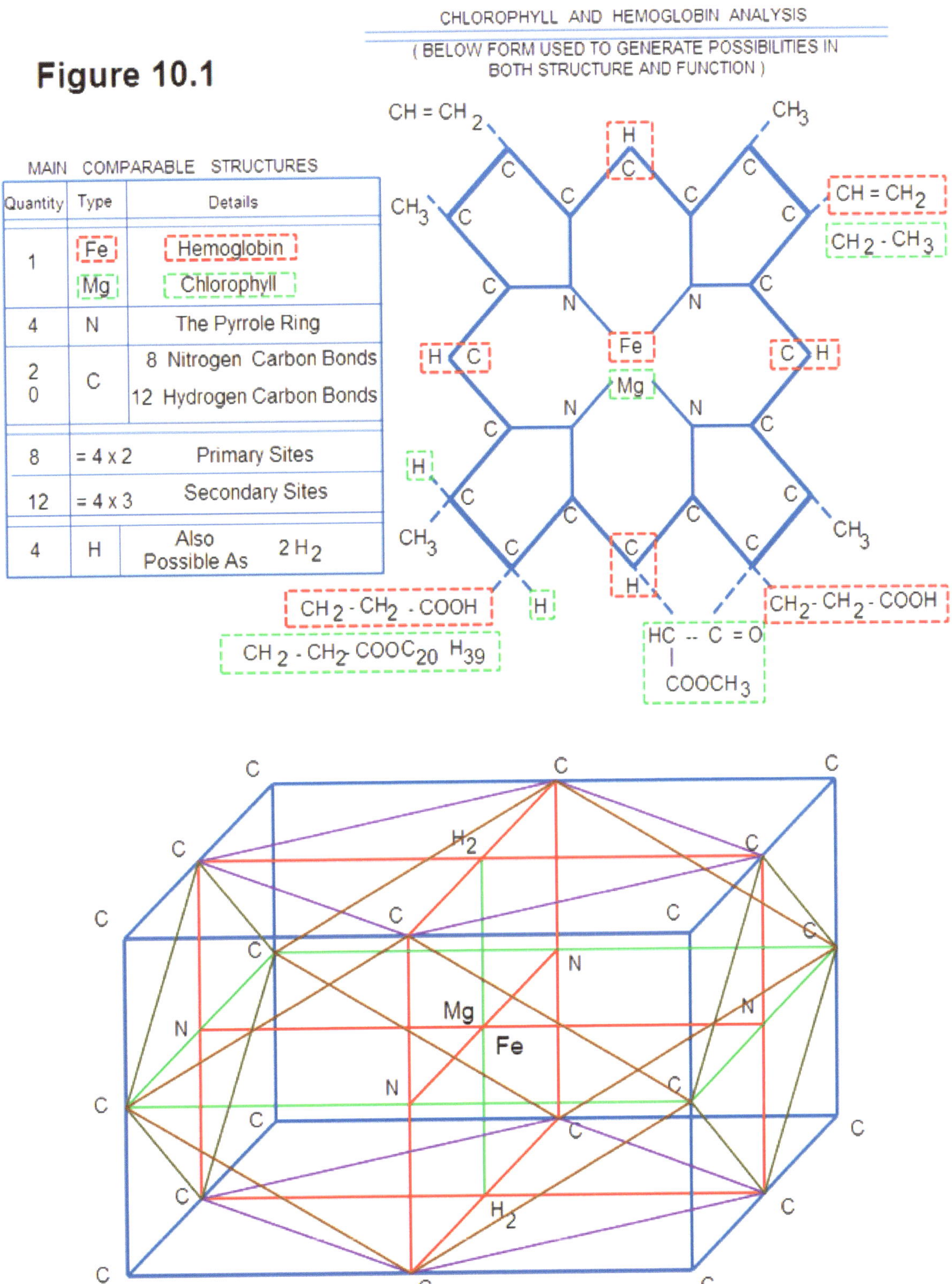

Figure 10.1

This uniquely remarkable existence of an elemental cycling within organic life seems to stem from the interrelationships in the molecular polarity of its counter parts. That is, a focal point of electromagnetic-like alignment presenting a characteristic of concentration accordingly to the existing pressures in part of its environment and the pertinacious particularity of its restive promiscuity. Porphyrin rings present one of the major organic building blocks for life on Earth. Within the heart of this molecular polarity, it would appear that plants and animals maintain a like vestige of this evolutionary bridge for growth and life. The only difference in their shared evolutionary process is rather the choice of metal used for the molecular polarity; iron for animal's hemoglobin versus magnesium for plant's chlorophyll. This very process that provides for enthalpic reactivity of their gaseous circulation becomes the life blood of animate life; oxygenation for animals versus photosynthesis for plants. Such becomes the primal phases of its existence and its survival to any and all forces within and without its medium. Such concentrations, seemingly attracting collection or solidification of deposits, where internal is always greater than external.

Yet while life shared this evolutionary process, the enigma of their elementary organic birth, subsistence, and divergence still remain as conjecture for discussion among theorists. The very nature of these two major evolutionary conditions embraces a dichotomy of organic purposes that maintains a primary symbiotic relationship: photosynthesis for oxygen and respiration for carbon dioxide. Still, it is the plant's chemical cycle which holds the greatest possibility of forming first as the postulate of life among organized being. Animal life would seem to have had its best probability to primordial promulgation upon the inversion of the plant's elemental cycle; a process necessary for the survival to a concerted effort in energy circulation within organic life. This is based on the expectation of a dense atmosphere of carbon dioxide persisting during the Earth's early development. While it is widely accepted that dissolved iron can chemically captured any free oxygen, the polaric relationship for the association of magnesium with carbon dioxide is not well understood. The theoretical consensus for this association began with the advent of photosynthesis by organisms (prokaryotic, then eukaryotic) that emitted oxygen as a waste product. Since carbon dioxide was plentiful and bondless oxygen was not, free oxygen was not a primary driver of organic life. As this waste product, free oxygen, eventually overwhelmed carbon dioxide, an oxygenated atmosphere gave rise to organisms that could chemically capture and capitalize on this free oxygen.

Eventually this process can be conjectured to have given rise to a whole host of functioning hybrids, some of which are now respectfully classified as neither plant nor animal. Other divergent life forms, such as mollusks and arthropods, are among those whose oxygenation is achieved in the respiratory function of hemocyanin. Instead of using iron for hemoglobin oxygen transport, these crustaceans living in low oxygen pressure use copper for hemocyanin oxygen transportation. Organic life appears to maintain a unique bond with some of the more electromotive series of metals: copper, iron, and magnesium. Interestingly, as well, blood pigment appears to mimic the corrosive coloring of its involved metal: chlorophyll is green, hemocyanin is blue, and hemoglobin is blue.

The diversity of aerobic and anaerobic organisms found only in the ocean's waters, lends credence to the hypothesis that virtually all such organized life can be deduced to have been originated in such an aqueous environment. Thus, it is believed that the evolution of such organized life is inexorably tied to the presence of large amounts of water together with an atmosphere of carbon dioxide high in nitrogen content. This would be little to no problem for a developing planet embryo maintaining the approximate distance from the sun as does the Earth. Subsequently first life would've been that area of water embodying some properties of osmosis to maintain its organic gaseous circulation for energy. Understanding the electrostatic properties of a bead of water retaining a regulative pressure to the competition ensued in oxidation versus reduction and subsequent enthalpies of reactants versus products, or heat versus cold, presents an interesting application of the preceding theory or primal life. Not quite considered a living entity as defined of science or religion, it would seem to implore the basic rudiments of pertinacious particularity and restive promiscuity.

Water, having an abnormally low vapor pressure, high boiling point, high melting point, high heat of fusion, and high heat of vaporization for such a small molecular weight, exists as the medium of expediency to the limited continuum of substance. This is, in part, due to its high molecularly bonded polarity; the lower the temperature, varying slightly with pressure, the higher its polarity. This bond is achieved upon the breakup of oxygen molecules by hydrogen molecules to which oxygen favors over its own bonded, polar, attraction. A highly exothermic reaction, this breaking of bonded oxygen molecules, its heat is just as quickly expanded, or absorbed, upon an equally reactive endothermic process of oxygen binding with hydrogen molecules. In terms of bond strengths within the molecules, an endothermic reaction indicates that the bonds are stronger in the reactants than in their products, while an exothermic reaction indicates that the bonds are stronger in the products than in the reactants. In this way, heat becomes the common osmotic denominator of evolution.

Yet, what is this substance or quality of heat? It might be said that it is a touch of warmth toward creativity for heat holds the electromotive force of creation itself. For where there is bonding, or building, toward something opposing an explication of nothingness, there is this specter of subsistence called heat. Even in the enigma of snowflake formation, the radial pattern of a magnetic influence of force, propagating the plant-like construction of a polar field in perpendicular to its bonding current, seems to predominate that aspect of confluence in the electromotive force wrought in the insistence of an induced polarity via its inertial transition. Thus, as the illusive spirit of life, its undulation permeates all subsistence with the quality 'to be.' But, still it does not relate the negativity of this opposition; rather it is a resultant character of its resistance to an acceleration in velocity. Herein inspiriting a deceleration of velocity, it enraptures solace with a perception of identity in measure of kinetic energy. Where non-existence is the expansive force of being, its acknowledgement is the contractive respite of entity in existence. Therein is it that a force of contraction can never completely displace a force of expansion least they both resign themselves to discontinuity by discharge. This principle is remedially distinguished upon the expectant annihilation of the properties defining the distinctiveness of a positron and an electron during just this type of incidental encounter; whereby the conjoining of these two parallel fields in perpendicular alignment form that ideorepulsive persistence of objectivity in contention, expressed as a photon. Indeed, this photon understands a subsistence of heat as the resultant character of resistance and, as such, therefore becomes that measure of energy perceived 'to be.'

Note, an electron passes from a high (kinetic) energy level (deceleration) to a low (potential) energy level (acceleration), the difference of energy is radiated as one photon, or quantum; the limiting or qualifying boundary of measure by which subjectivity perceives identity; or those characteristics enlivening our sensibility 'to be.' Such is the field of influence to opposing forces in polaric complement that acknowledges the relativity of existence as a perception of difference. See causes and results for these types of transitions below:

If Expanding mass condenses resistance
There is an excess of neutrons and the matter is considered radioactive;

As Condensing resistance expands energy
And Expanding energy condenses matter
There is heat of fission from an exothermic reaction, or loss of an electron. The neutron disintegrates for more energy and less total mass;

If Condensing matter expands resistance
There is an excess of protons and the matter is considered transmutative;

As Expanding resistance condenses energy
And Condensing energy expands mass
There is heat of fusion from an endothermic reaction, or addition of an electron (often thought as a loss of a positron). The proton disintegrates for forced formation of a neutron for less energy and greater total mass.

Fusion, as a continuum in convection of perpetual motion, does not necessarily have to assimilate the countenance of large bodies like stars or black holes. Resistance is then understood as the balance between energy and mass whereby heat becomes the impression, or expression, in its detention of less energy in its desire for conjugation: An 'intinction' (or consecration) of corporal convection in the osmotic equalization of enthalpies in energy. To understand energy is to comprehend the resilient nature of resistance. To understand electricity is to comprehend directional attitudinizing of resistance as the ability to achieve parallel alignment of a perpendicular expectant. That is, the retention of transmigration via polaric induction. Herein negative pressure, or cold, absorbs a positive pressure, or heat, in a moiety of attraction to a mutual immediacy for equilibrium: This is entropic equanimity. Yet, even as the positron is reflected as a loss of an electron, so is it that it shall be held in a contention of indecision upon its loss from origin, as well as its loss to any relocation. Like a stretching rubber band, the transient nature of the positron field is held in retention of its transmigration in contraction of, or packing of, an electron field within the medium of conductivity abetting resistance; especially on consideration of mass as that character of non-existence existing within the limitations of its own identity to perception: Matter

What has been related, in this examination of heat and resistance, is the induced field of positive pressure upon the resistance of electron flow through a medium of forced polarity. This field is similarly comparable, in design, to the conductivity of the nervous system of our own bodies in that the cylinder, or tube, formed by increased ordering of electrons within this field of influence, accentuates, via parity, the increased disordering of electrons surrounding this field. This separation of electrons into diverse entropies demonstrates a repulsion of those free electrons, for a more positive field, from an exaggerated electro-negativity of electron packing induced as current within a field of forced polarity. Obviously the greater pull, or the electron field's affinity for a positron field, the greater the induced field of positive pressure through a medium of resistance and the greater the probability of breaking weaker bonds of polarity which generate heat as fusion, or perpendicular expectant, of two opposing charges, or reciprocal forces, sometimes forming compounds in their wake.

Such fields of forced polarity can be induced and therefore demonstrated in many ways and forms: i.e., the appearance of lightning striking, the induced magnetism of certain metal elements upon striking, the striking of almost any object resulting in material separation in the form of cracks. Indeed, what we consider as our ability to break, tear, or otherwise destroy is merely our ability to induce a forced polarity within the affected mass. When striking a stone or glass, a forced polaric reaction, similar to that of lightning, instantaneously pits free electrons against each other along a conduit of current within the stone glass molecules. In this way, the bonding has been altered by an application of energy in the form of this striking force. This is less apparent and not as instaneous in other materials. In separating metal, this ability to induce forced polarity is not as instaneous. Metal is bent back and forth, creating heat of resistance, to induce segmented breaks in the mass where free electrons have gathered to repel and separate the metal molecules. Perhaps this concept of free electrons massing together under heat and pressure could even be used to better predict metal fatigue or even impending earthquakes.

Closing Statements – Sentient Matter

Herein illustrated, oftentimes connotatively, are energy, motion, and matter. What has not been presented heretofore is the proper physical 'cause and effect' relationship of individual pursuance. It is quite simple to state – first there is plant life, then a remedial hybrid, and finally animal life; a logical conjecture forming the basis of the evolutionary process, to be sure, but it does not give us the concept of independent motive from a ganglia of collective energy stores imposing perception into motion. Where does just an automatic function of simple reflex perfunctory mechanics leave off and give rise to an autonomic function of motive force in a modicum of rationality qualified via desire. This is the query to which theologians should be proposing and utilizing to pontificate a remonstration of the secular view to an evolutionary process, if it were possible. This is the real modulator of mystical intervention incurred even before the inimitable initiation of an articulate appreciation of thought and reason.

Fortunately, the mystery is not all that preclusive. All life acknowledges and perceives energy via motion in velocity. Occlusions of such subsistence are brought into motion in reaction to its environment; kind of an osmotically enthalpic reactivity. Although, as a register of these impressions, how does an entity utilize a need without external prodding by 'Mother Nature?' Again, how could this osmotically enthalpic reactivity presume to contrive a motive of imposition for any of our five senses of perception? Think about it: What unit of life was first to take in food, move of its own accord, attain the qualities necessary to be said of it, "There it is..., animal life; see how it fends for itself." As one can see upon review, I am not talking about reason by choice, rather I am investing a responsiveness in thought via need; that is, 'it does' to survive or it perishes (similar to Charles Darwin's concept of natural selection, promoted by his work, 'The origin of Species'). If it perishes, then it does not pass on any ability to be retained. If 'it does', managing to survive, then it does pass on an ability to be retained by its offspring (or mitotic replication). And this retained ability is added to the sequence of its procreation to be played out during its gestational development.

Much like a computerized conjuncture, it relates a memory of programmed responses via experience; if it is not needed, it does not get done. So to a question one might ask, who is the master programmer? The answer is that articulation of evolution expressed as God, for the being of energy/existence is certainly dependent upon non-existence. But to a question one might ask, how is it then programmed? The answer is by its being, existing, and registering a perception of its function from its perfunctory response to its environment. For, as long as there is an imbalance of polaric division in kinetic and potential forms, energy will continue to outline an eighth (or probably tenth) dimensional continuum of matter with a motive to existence.

Therefore one is seeking, as resolution, a forced polarized induction for independent motive force via acknowledgement or learned reactivity as a necessity or capability of doing useful work. Such is a characterization of entropic equanimity in the free energy change of enthalpic reactivity. Free energy is responsible for the spontaneity of a reverse reaction, or reciprocal response, in the nature of projecting memory programmed perceptions as the indirect stimulus of motive force. Free energies are stored, impressed, in endothermic configurations of perpendicular constraint and are released in exothermic reactions of parallel, or polaric, alignment. These reactions, induced upon the learned equivalent response of desire for entropic equanimity, survival, perceived in a unit of life, autonomously express the necessary impressions needed to maintain identity in being: Sort of a self-maintenance program.

These reactions are then indirectly related to an articulation of its acculturation to its environment, suspended by some half-life inculcation to a reverse reaction time as opposed to a direct relation of spontaneity. It is here that memory is infused into the ganglia of life's essence and thought proliferates in an associative assembly of free energy, inevitably to foster an increased potential demonstrating 'free will.' Such is it that we promulgate the level of an entity's intelligence by how much memory it can draw upon at will for associative assembly of thought toward its ability to enact upon or communicate its desires. Consequently this compartmentalizing of free energies enlivens a contrariety of 'being' that inspirits an insanity to be ordered within the chaos of an entity's capability for intelligence. This ordering provides for the random ability to create and revise connections within the multitude of available memory as a way to more efficiently process its needs and desires. This insanity gives way to hallucinations, imagination, and the ability to abstract that which is not real. Out of this proclivity for order, 'free will' conceives its moral dogma for right and wrong. It is a process by which the mental circulatory process mimics its origin to gain balance. As existence both repels and attracts non-existence for being, 'free will' attempts to define truth from fiction, desires from more practical needs, and right from wrong. So in a sense, we are all born insane due to our inheritance of 'free will'.

And yet, while the structure of the brain may be inheritable from one generation to the next, a precarious variety of other chemical and environmental events can alter its blueprint of connections and ultimate development. Still the primary function of such organized memory, as retained and passed on, is to act as the governor of every ongoing event within the organism that keeps its motive, alive. The more complex the organism, the more of the memory is required for such maintenance. The human brain consumes up to twenty percent of the energy used by the human body, more than any other organ. It has been rumored that the humans only use some small percentage of their brain. This is not based on any real science and is probably a hold over from misunderstandings (or misrepresentations) of neurological research conducted in the late 19th century or early 20th century, in which they physically invaded the brain to study affected behaviors; unbelievably primitive scientific behavior at best. The belief was that more of the brain could be dedicated to increase one's intelligence. Rather intelligence can only be increased by the ability to make impressions and associative connections that would allow for more efficient retrieval of these memory stores. Consequently, each progeny is provided with a structural blueprint and associated connections consistent with their evolution. However the exercising or reinforcing of these connections is still left to the individual organism.

Here, one sees the major constituents of organic life as the incipient character of energy innate to all matter; science provides us with a calculated construction of this understanding. Here, one sees evolution as the furthered increase of positive entropy seeking negative entropy for equanimity; religion provides us with a ratiocinated construction of this truth to be taken on faith. Faith, trust, and hope are the resultant products of our ability for abstraction in an effort to mimic a potential insistence to bring about a kinetic persistence. Here, one asks, will a continued evolution breed a greater insanity in a greater capacity to understand that which defies explanation? Herein lays the desideratum of kinship between science and religion. For truth cannot be upset by the pursuance of knowledge, understanding, there is a need and a place for both concernments. But truth and knowledge can not be given, they must be won; indeed they must be 'one' again.

Therein lies still another dimension: It is a dimension in knowledge as the omniscient quality of relativity that satisfies the enticement of kinetic procreation. It is weighed and bounded in the synteresis of relativity and supersedes all else. It is the 'see' of perdition and the fruit of pansophy, culminating as the crucible of pantheism proposed in truth. It is the truth proposed in the ratiocination of a precision in science and a precisian in religion in which it can truly be said: "Sic itur ad astra"; such is the way to immortality! If this seems a peculiar end, your right – this is only our beginning to a new understanding of the life and the universe.

www.ingramcontent.com/pod-product-compliance
Ingram Content Group UK Ltd.
Pitfield, Milton Keynes, MK11 3LW, UK
UKHW060120300726
14090UKWH00002B/283

* 9 7 8 1 4 6 2 8 8 7 5 9 0 *